# Understanding Electromagnetic Scattering Using the Moment Method

## A Practical Approach

Randy Bancroft

**Software to accompany this book is available for download at:**
http://www.artech-house.com/static/bancroft859.html

Artech House
Boston • London

**British Library Cataloguing in Publication Data**
Bancroft, Randy
 Understanding electromagnetic scattering using the Moment Method : a practical approach
 1. Electromagnetic waves — Scattering 2. Electromagnetic waves — Scattering—Data processing
 I.Title
 539.2'0285

ISBN 0-89006-859-3

© **1996 ARTECH HOUSE, INC.**
**685 Canton Street**
**Norwood, MA 02062**

International Standard Book Number: 0-89006-859-3

10  9  8  7  6  5  4  3  2  1

To the muses who inspired two degrees:

Dawn Herman & April Masché

And a mentor who provided encouragement:

Richard Booton

# Contents

Preface                                                                         ix
Chapter 1  Numerical Integration                                                 1
  1.1  Singular Integrands                                             1
  1.2  Richardson's Extrapolation                                      4
  1.3  Midpoint Integration With Richardson's Extrapolation            7
  References                                                          11

Chapter 2  Moment Method                                                        13
  2.1  Surface Charge on a Conductive Strip                            13
  2.2  Galerkin's Method                                              20
  2.3  Symmetry                                                       22
  2.4  Capacitance of a Square Conducting Plate                       24
  2.5  Concluding Remarks                                             31
  References                                                          33

Chapter 3  Thin Wire Scattering                                                 35
  3.1  Hallen's Equation                                              35
  3.2  Moment Method Solution (Pulse/Delta)                           38
    3.2.1  Computational Results (Pulse/Delta)              42
  3.3  Moment Method Solution (Triangle/Delta)                        46
    3.3.1  Computational Results                            48
  3.4  Moment Method Solution at Arbitrary Incidence                  54
  References                                                          59

Chapter 4  Scattering From Conductive Strips                                    61
  4.1  RCS of Perfectly Conducting Strip                              61
    4.1.1  TM Polarization                                  61
    4.1.2  TE Polarization                                  64
  4.2  Calculation of Two–Dimensional RCS                            70
  4.3  Numerical Results of TM and TE Scattering                      71
  4.4  TM Scattering From Resistive Strips                            71
    4.4.1  Quadratic Resistive Taper                        75
    4.4.2  Taylor Resistive Taper                           77
  References                                                          80

Chapter 5 Scattering From Two–Dimensional Contours     81
  5.1 RCS of Perfectly Conducting Two–Dimensional Contour     81
    5.1.1 TM Polarization     81
    5.1.2 Numerical Results for TM Scattering     84
    5.1.3 TE Polarization     86
    5.1.4 Calculation of RCS in the TE Case     90
    5.1.5 Numerical Results of TE Scattering From a Contour     94
  5.2 Monostatic and Bistatic RCS     94
  References     98

Chapter 6 Radar Cross Section of a Flat Plate     99
  6.1 RCS of a Thin, Perfectly Conducting Square Plate     99
    6.1.1 Moment Method Solution (Pulse/Pulse)     99
    6.1.2 Moment Method Solution (Rooftop/Pulse)     113
    6.1.3 Numerical Results     117
  6.2 Concluding Remarks     123
  References     124

Appendixes: FORTRAN Computer Programs     125
  Appendix A: Chapter 2 FORTRAN Computer Programs     127
  Appendix B: Chapter 3 FORTRAN Computer Programs     147
  Appendix C: Chapter 4 FORTRAN Computer Programs     163
  Appendix D: Chapter 5 FORTRAN Computer Programs     193
  Appendix E: Chapter 6 FORTRAN Computer Programs     229

Index     253

# Preface

In the late 1970s when I first took an interest in electromagnetics (EM), my university professors encouraged the use of computers. They stressed that while many problems in EM have no closed–form solution, they could be solved using computer methods. The use of the Fast Fourier Transform and some Finite Difference calculations were introduced in our coursework, and use of numerical methods caught my fancy.

When I took a class in antenna theory, the textbook's chapter on moment methods was omitted. My interest in numerical methods drew me back to this chapter. When I read it, I found the material as presented confusing. Analogies with Kirchoff's network equations did not enlighten but seemed to obscure an understanding of this method. The following sections of the textbook contained mathematics of a much higher level. I worked through some of the examples, but with the limited computer resources available to me at the time, I was discouraged.

I purchased another work on using the moment method to solve EM problems. It spent considerable time discussing linear spaces, before solving problems with the moment method. This led me to the mistaken belief that I could not hope to understand the moment method without first understanding the intricacies of linear spaces.

By the late 1980s personal computers became powerful enough to solve many introductory moment method problems with ease. A number of my colleagues often asked me for examples of the moment method to use as a guide in learning how to apply the method. I have written this book as an easy introduction on the use of the Moment Method. The examples contained will run on most personal computers (486DX–33 or faster, compiled using Microsoft FORTRAN, Watcom or Lahey). This book is a practical guide to teach the reader to use the moment method with as little mathematical formalism as possible. I chose various examples so the reader will develop a feeling for the currents generated by electromagnetic waves on a wire, strip, or plate. The advantages of using extrapolation is also prominently illustrated.

A few years ago, Professor Robert E. Collin related what he sees as a problem with common practices in computational EM:[1]

---

[1] Robert E. Collin, "The Role of Analysis in an Age of Computers: View From the Analytical Side," *IEEE Antennas and Propagation Magazine*, August 1990, pg. 27–31.

... Authors have generally been more interested in the formulation of a problem for numerical solution, than in generating accurate numerical results. In many published papers, authors do not document studies carried out on the numerical convergence, or any assessment of the absolute accuracy achieved for their numerical results. One frequently finds that authors do not record how many basis functions were used, how the numerical results would change if the number of basis functions were doubled, how many basis functions must be used to achieve a numerical convergence to within 1 percent, or any checks on absolute convergence to the correct answer. The user is left with the burden of trying to assess the reliability of the published results, and without sufficient information to repeat the numerical evaluation.

A good paper on a numerical solution to a particular problem should have many of the attributes of a good paper describing an experimental investigation. Sufficient information should be provided so that others can repeat the numerical experiment, along with a careful evaluation of the expected accuracy of the results presented.

The format of this book may strike some as excessively detailed. This is a deliberate choice. It is in the spirit that Professor Collin expressed so eloquently above.

I would like to thank Richard Booton for his patient discussions of the problems in this work, and for teaching me the basics of the popular numerical methods used in electromagnetic analysis.

# Chapter 1

# Numerical Integration

## 1.1 SINGULAR INTEGRANDS

Numerical integration is seldom discussed in texts that concern the moment method. Authors generally assume that their readers are knowledgeable concerning the subject. As electromagnetics books have a review section for vector analysis, it makes sense that a book on numerical electromagnetics could spend a few pages discussing numerical integration.

The moment method converts an equation that contains a linear operator into a matrix equation. Often the operator is an integral, which just as often contains an integrand with a singularity. The integrand may have a singularity at one (or both) of the integration limits, or one may exist within the integration interval.

Popular algorithms that approximate an integration are often referred to as *open* and *closed*. If the algorithm is open, the integrand is not evaluated at the integral's limits (i.e., at the endpoints). A closed integration formula uses the integrand evaluated at the integration limits.

*Simpson's rule* and *Gaussian quadrature* are two very popular integration methods. They rely on the assumption that the integrand may be conveniently approximated with a polynomial. When this assumption is violated, these formulas can produce inaccurate or even completely misleading results. Poor results can also occur if an integrand has singularities in its derivatives. This is less serious but can cause significant error. There are a number of methods that can tame an unruly integral. A few of them are summarized below:

1. Ignoring the singularity may sometimes prove a successful approach. A polynomial–based integration method may converge as the integration interval is partitioned into smaller and smaller subsections.

2. Series expansions of an integrand may sometimes be integrated term by term. This clearly works best when the series converges quickly.

1

3.  Subtracting the singularity is a procedure in which an integral may be split in
    two. One term contains the singularity and is amenable to analytic solution.
    The other term contains a smooth integrand that is readily integrable using
    polynomial methods.

4.  Change of argument is a very successful method. An integral may be rewritten
    into a form with less severe singularity, or in some cases the singularity may be
    completely removed.

5.  Gaussian methods have been developed to deal powerfully with certain singu-
    larities.

Illustrations of these techniques exist in technical literature.[1] [2] For the purposes of this work, we chose a rough–and–ready approach to numerical integration.
A very simple open integration formula is to evaluate the integrand at a number
of points that correspond to the center of a subinterval, add these values together,
and multiply by the subinterval length $H$. By assuming a constant length for each
subinterval, we may use the single constant $H$, which simplifies the integration con-
siderably. This integration scheme is expressed mathematically in equation (1.1).
This situation is illustrated in Figure 1.1.

$$\int_a^b f(x)\, dx \approx H \cdot \sum_{n=1}^{N} f((n-1/2)H) \qquad (1.1)$$

$$H = \frac{b-a}{N} \qquad (1.2)$$

This situation produces an integration scheme that does not evaluate the integrand at the integration interval's endpoints. Should a singularity exist at an
endpoint, it is excluded and will not cause a computer overflow. We refer to this
method as *midpoint* integration.

When we implement the moment method, a second variable in the integrand,
which is not a variable of integration, is set to the value at the center of an integration interval. This is often the case when we implement a moment method solution
called *point matching*.

If the subintervals for midpoint integration divide the integration interval using
a multiple of two (e.g. 2, 4, 8, etc.), this division of the integration interval avoids
evaluation of the integrand at the subinterval center, which is often a singularity.

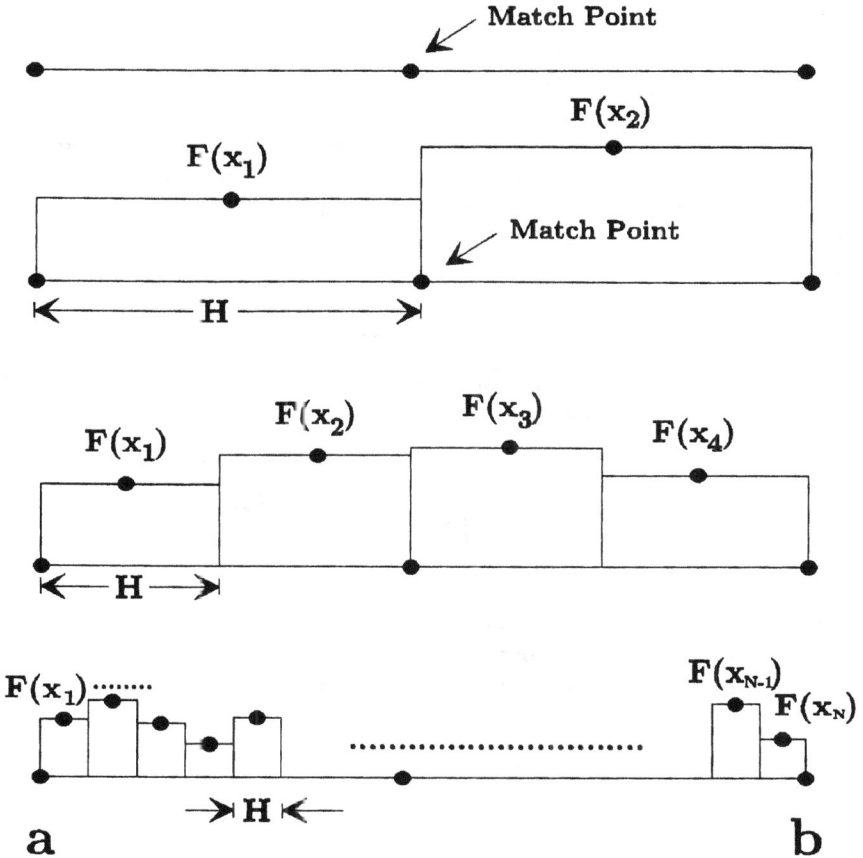

**Figure 1.1** The top illustration is an arbitrary integration interval $[a, b]$. The match point (singularity point) is at the center, it is a value that we do not want to evaluate when estimating the integral. We next divide the interval into two segments. The integrand may be evaluated at each of these two points, added, and multiplied by $H$ (the subinterval length) for an integration estimate. These subintervals can be halved and the process repeated. The last line illustrates the process in general.

We have thus developed an integration scheme that avoids common singularities in moment method implementations. Unfortunately, this method often converges very slowly. In general so slowly, that round–off error begins to contaminate the integration results before we obtain a useful estimate. To improve this situation, we explore a method known as *Richardson's extrapolation.*

## 1.2 RICHARDSON'S EXTRAPOLATION

Richardson's extrapolation is a method which can take a lower order estimation and produce a higher order estimate.[3] [4]  This extrapolation technique assumes that we have a formula $E(H)$ which produces approximations of error $O(H^2)$ to an unknown value $Q$. We also assume the error form for the approximation of $E(H)$ to $Q$ is:

$$Q = E(H) + K_1 H^2 + O(H^4) \qquad (1.3)$$

$K_1$ is a constant which is independent of $H$. If we halve the value of $H$ and place it into (1.3) we obtain:

$$Q = E\left(\frac{H}{2}\right) + K_1 \frac{H^2}{4} + O\left(\left(\frac{H}{2}\right)^4\right) \qquad (1.4)$$

We note that the order of accuracy of (1.4) is still $O(H^4)$. Equations (1.3) and (1.4) form a set of simultaneous equations that allow us to eliminate $K_1$ and solve for our desired value $Q$. We multiply (1.4) by 4 and subtract (1.3):

$$3Q = 4E\left(\frac{H}{2}\right) - E(H) + O(H^4) \qquad (1.5)$$

Dividing by 3 we obtain a value for $Q$:

$$Q = \frac{4E\left(\frac{H}{2}\right) - E(H)}{3} + O(H^4) \qquad (1.6)$$

We could next postulate that a number $K_2$ exists which allows us to write (1.3) as:

$$Q = E(H) + K_1 H^2 + K_2 H^4 + O(H^6) \qquad (1.7)$$

We can halve the interval once more and solve for an $O(H^6)$ estimate of $Q$. In general if the error form of $E$ may be expressed as:

$$Q = E(H) + \sum_{i=1}^{m-1} K_i H^{2i} + O(H^{2m}) \tag{1.8}$$

Then $O(H^{2i})$ approximations are generated recursively by:

$$\bar{E}_i(H) = \frac{4^{i-1} E_{i-1}(H/2) - N_{i-1}(H)}{4^{i-1} - 1} \tag{1.9}$$

for each $i = 2, 3, \ldots, m$.

We are now ready to put Richardson's extrapolation to work in an example. Consider the function:

$$F(x) = x^3 \sin x \tag{1.10}$$

An approximation for the derivative of a function is:

$$\frac{dF}{dx} \approx \frac{1}{2H}[F(x+H) - F(x-H)] \tag{1.11}$$

The exact expression for the derivative of (1.10) is:

$$\frac{dF}{dx} = 3x^2 \sin x + x^3 \cos x \tag{1.12}$$

If we choose $x_0 = \pi/4$ as a point where we desire the derivative of (1.10), we can evaluate it using (1.12):

$$\frac{dF(x_0)}{dx} = 1.6511112328$$

We now use (1.11) with $H$ arbitrarily chosen as 0.1 for the initial estimate:

$$E(H) = E(0.1) = \frac{1}{2(0.1)}[0.53734170 - 0.20380728] = 1.66767213$$

We halve $H$ and obtain a second estimate:

$$E(H/2) = \frac{1}{2(0.05)}[0.43234357 - 0.26681735] = 1.65526221$$

Halving $H$ again we obtain a third estimate:

$$E(H/4) = \frac{1}{2(0.025)}[0.38562989 - 0.30302241] = 1.65214965$$

In the limit as $H \to 0$ in (1.11) the value of the function becomes the exact derivative. We can see that for our values of $H$ we obtain approximations which in the case of $E(H/4)$ is only good to two places past the decimal. Applying Richardson's extrapolation:

$$E_1(H) = 1.66767213$$

$$E_1(H/2) = 1.65526221$$

$$E_2(H) = \frac{4(1.65526221) - 1.6676721}{3} = 1.65112558$$

$$E_1(H/4) = 1.65214965$$

$$E_2(H/2) = \frac{4(1.65214965) - 1.6552622}{3} = 1.65111213$$

$$E_3(H/4) = \frac{16(1.65111213) - 1.65112558}{15} = 1.65111124$$

We now have an estimate good to seven places past the decimal point.

## 1.3  MIDPOINT INTEGRATION WITH RICHARDSON'S EXTRAPOLATION

We may now use Richardson's extrapolation with midpoint integration to obtain an accurate estimate of an integral. We begin by choosing $N = 2^m$ for equation (1.2). This will halve our integration interval for $m = 1, 2, 3, \ldots, M$. We may then use extrapolation to obtain higher order estimates for the integral.

**Table 1.1**

| M | Midpoint Integration | Richardson's Extrapolation |
|---|---|---|
| 2 | 1.577350 | 1.577350 |
| 4 | 1.698844 | 1.820338 |
| 8 | 1.786461 | 1.891991 |
| 16 | 1.848857 | 1.928165 |
| 32 | 1.893088 | 1.950607 |
| 64 | 1.924392 | 1.965540 |
| 128 | 1.946535 | 1.975794 |
| 256 | 1.962195 | 1.982941 |
| 512 | 1.973266 | 1.987952 |
| 1024 | 1.981095 | 1.991487 |
| 2048 | 1.986635 | 1.993998 |
| 4096 | 1.990546 | 1.995730 |
| 8192 | 1.993318 | 1.997011 |
| 16384 | 1.995274 | 1.997863 |
| 32768 | 1.996658 | 1.998500 |
| 65536 | 1.997631 | 1.998915 |

To illustrate the use of midpoint integration with Richardson's extrapolation, we use (1.13):

$$\int_0^1 \frac{1}{\sqrt{x}} \, dx = 2 \qquad (1.13)$$

Table 1.1 presents the results for midpoint integration alone and Richardson's extrapolation of the data. Note that we automatically avoided the singularity at the lower limit using this integration scheme. The midpoint integration alone has an error of $2.369 \cdot 10^{-3}$, where the small algebraic effort of Richardson's extrapolation improves this to $1.085 \cdot 10^{-3}$.

The FORTRAN code that produced Table 1.1 is included below:

```
C
C
C     ****************************************************************
C     ** THIS SUBROUTINE APPROXIMATES THE INTEGRAL OF A FUNCTION  **
C     ** USING MIDPOINT INTEGRATION WITH RICHARDSON EXTRAPOLATION **
C     ****************************************************************
C

      IMPLICIT NONE
      REAL CP(16,16),C1,SUMX,MIDINT
      REAL NSX,DX,A,B,MIDPNT,ERROR,OLD,NEW,TOL,PI
      INTEGER COUNTX,I,Q,NMAX
      LOGICAL FINISHED
C
      FINISHED=.FALSE.
      COUNTX=1
      TOL=5.0E-7
      NMAX=16
      PI=ACOS(-1.0)
C
C     *****************************************
C     ** LOWER AND UPPER LIMITS ON INTEGRAL **
C     *****************************************
C
      A=0
      B=1
C
      DO WHILE(.NOT.FINISHED)
C
C        *************************************
C        ** CALCULATE NUMBER OF SUBSECTIONS **
C        *************************************
         NSX=2.0**COUNTX
C
C        *************************
C        ** CALCULATE STEP-SIZE **
C        *************************
         DX=(B-A)/NSX
C
C        ***********************************************
C        ** EVALUATE AND SUM FUNCTION VALUE AT MIDPOINTS **
C        ***********************************************
C
         SUMX=0.0
C
         DO 10 I=1,NSX
            MIDPNT=A+FLOAT(I-1)*DX+DX/2.0
            SUMX=SUMX+1.0/SQRT(MIDPNT)
 10      CONTINUE
C
C
```

```
C     **************************************************
C     ** ASSIGN SUMATION TO ARRAY FOR EXTRAPOLATION **
C     **************************************************
      CP(1,COUNTX)=SUMX*DX
C
C     *********************************************
C     ** RICHARDSON EXTRAPOLATION CALCULATION **
C     *********************************************
C
      C1=1.0
C
      DO 20 Q=2,COUNTX
        C1=2*C1
        CP(Q,COUNTX)=(C1*CP(Q-1,COUNTX)-CP(Q-1,COUNTX-1))/(C1-1.0)
 20   CONTINUE
C
      MIDINT=CP(COUNTX,COUNTX)
C
      NEW=CP(COUNTX,CCUNTX)
C
C     **************************************
C     ** CALCULATE ERROR AFTER FIRST PASS **
C     **************************************
C
      IF(COUNTX.GT.1) ERROR=ABS(OLD-NEW)/ABS(NEW)
C
C     ****************************************************************
C     ** CHECK ERROR AFTER FIRST PASS TO SEE IF INTEGRATION   **
C     ** IS WITHIN TOLERANCE END CALCULATION IF IT IS         **
C     ****************************************************************
C
      IF((COUNTX.GT.1).AND.(ERROR.LT.TOL)) FINISHED=.TRUE.
C
C     ****************************************************************
C     ** CHECK TO SEE IF WE HAVE REACHED THE MAXIMUM NUMBER **
C     ** OF SUBSECTIONS ALLOWED BY USER                     **
C     ****************************************************************
C
      IF(COUNTX.EQ.NMAX) FINISHED=.TRUE.
C
      COUNTX=COUNTX+1
C
      OLD=NEW
C
      WRITE(*,*)NSX,SUMX*DX,MIDINT
C
      END DO
C
C
      RETURN
      END
```

One can encounter cases in which this method breaks down or does not converge, but like a jeep, midpoint integration with Richardson's extrapolation covers a lot of mathematical terrain.

It is possible to present integrands that will cause any of the popular integration methods to produce incorrect or poor results. The reader is cautioned to keep in mind the importance of making certain that the integration scheme used is converging.

# References

[1] Scheid, Francis, *Numerical Analysis (Schaum's Outline Series),* New York, McGraw-Hill, 1968, chapter 16

[2] Zwillinger, Daniel, *Handbook of Integration,* Boston, Jones & Bartlett Publishers, 1992, chapter 5-6

[3] Burden, Richard L. and Faires, Douglas J., *Numerical Analysis,* fourth edition, Boston, PWS–Kent Publishing, 1989, pg. 157-163

[4] Richardson, L. F. and Gaunt, J. A., "The Deferred Approach to the Limit," *Philosophical Transactions of the Royal Society of London,* **226A**, 1927, pg. 299-361

# Chapter 2

# Moment Method

## 2.1 SURFACE CHARGE ON A CONDUCTIVE STRIP

The first problem we will address using the moment method is the determination of charge distribution across a conductive strip. In Figure 2.1 we have a metallic strip that is very long in the $\pm z$ direction. Our task is to determine the charge distribution along the $x$–axis.

We can imagine the strip to consist of line charges stacked next to one another and parallel to the $z$–axis. The potential ($\phi$) at a field point $\vec{\rho}$ produced by a line charge (electric field source) of density $q_L$ is given by:

$$d\phi(\vec{\rho}) = \frac{q_L}{2\pi\epsilon_0} \ln\left(\frac{1}{|\vec{\rho} - \vec{\acute{\rho}}|}\right) \tag{2.1}$$

Where:
$\epsilon_0 = 8.854223 \cdot 10^{-12}$ Farads/meter
$\vec{\rho} = $ location at which we observe potential
$\vec{\acute{\rho}} = $ source (i.e., charge) location

The surface charge function is constant along $z$ at any $x$–axis point. We can relate the surface charge on a strip to a line charge by multiplying the surface charge density by a differential change in $x$:

$$q_L(\acute{x}) = q_s(\acute{x})d\acute{x} \tag{2.2}$$

Placing (2.2) into (2.1):

$$d\phi(\vec{\rho}) = \frac{q_s(\acute{x})}{2\pi\epsilon_0} \ln\left(\frac{1}{|\vec{\rho} - \vec{\acute{\rho}}|}\right) d\acute{x} \tag{2.3}$$

Integrating both sides:

$$\phi(\vec{\rho}) = \int_{-a/2}^{a/2} \frac{q_s(\acute{x})}{2\pi\epsilon_0} \ln\left(\frac{1}{|\vec{\rho} - \vec{\acute{\rho}}|}\right) d\acute{x} \tag{2.4}$$

13

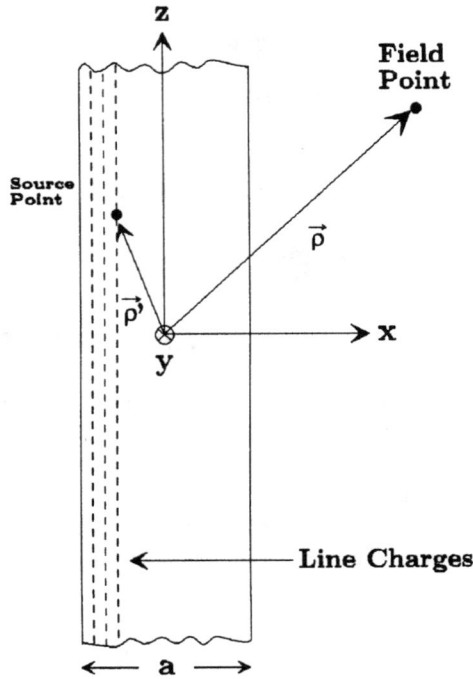

**Figure 2.1** A metal strip of width $a$ contains a surface charge that produces a constant potential $\phi$ at its surface. The strip is assumed to be very long in the $\pm z$ direction. We are interested in determining the charge distribution across the strip (i.e., along the $x$–axis).

This gives us the potential produced by the strip charge along the $z$ direction. We can now move the place where we observe the potential (point $P$) to the surface of the strip and confine the charge and observation points to the $x$–axis.

$$\phi(x) = \int_{-a/2}^{a/2} \frac{q_s(\acute{x})}{2\pi\epsilon_0} \ln\left(\frac{1}{|\,x - \acute{x}\,|}\right) d\acute{x} \tag{2.5}$$

We choose 1 V as the potential produced by the surface charge of the strip. For convenience, the width of the strip is set to $a = 2$. With these assumptions, (2.5) becomes:

$$\int_{-1}^{1} \frac{q_s(\acute{x})}{2\pi\epsilon_0} \ln\left(\frac{1}{|\,x - \acute{x}\,|}\right) d\acute{x} = 1 \tag{2.6}$$

We want to determine $q_s(\acute{x})$. We can choose to observe the potential at any point along $x$ we desire. If we choose $x = 0$, (2.6) becomes:

$$\int_{-1}^{1} \frac{q_s(\acute{x})}{2\pi\epsilon_0} \ln\left(\frac{1}{|\acute{x}|}\right) d\acute{x} = 1 \qquad (2.7)$$

This integral has the same form as a known definite integral:

$$\int_{-1}^{1} \frac{1}{\sqrt{1-\acute{x}^2}} \ln\left(\frac{1}{|\acute{x}|}\right) d\acute{x} = \pi \ln 2 \qquad (2.8)$$

normalizing the right–hand side:

$$\int_{-1}^{1} \frac{1}{\pi \ln 2\sqrt{1-\acute{x}^2}} \ln\left(\frac{1}{|\acute{x}|}\right) d\acute{x} = 1 \qquad (2.9)$$

Comparing (2.7) and (2.9) we conclude:

$$\frac{q_s(\acute{x})}{2\pi\epsilon_0} = \frac{1}{\pi \ln 2\sqrt{1-\acute{x}^2}} \qquad (2.10)$$

Rearranging (2.10), we obtain the exact solution of the charge distribution as a function of $x$:

$$q_s(x) = \frac{2\epsilon_0}{\ln 2\sqrt{1-x^2}} \qquad (2.11)$$

We now use the moment method to solve (2.7) with $x \neq 0$. Throughout this book we will encounter integral equations of this form:

$$\int_{a}^{b} h(\acute{x})G(x,\acute{x})d\acute{x} = g(x) \qquad (2.12)$$

In our case:

$$h(\acute{x}) = q_s(\acute{x}) \qquad (2.13a)$$

$$G(x,\acute{x}) = \ln\left(\frac{1}{|x-\acute{x}|}\right) \qquad (2.13b)$$

$G(x,\acute{x})$ is the kernel or Green's function of the integral equation. A good introduction to the concept of a Green's function as used in EM is presented by Sadiku.[1] It usually contains a pair of variables. The primed variable is the location of the source $h(\acute{x})$, which produces the observable quantity $g(x)$ located at $x$. In our case, the observed quantity is potential and the unknown quantity is charge/unit length:

$$g(x) = \phi(x) \qquad (2.14)$$

The implementation of the moment method begins by introducing a summation of constants multiplied by assumed known functions $f_n(\acute{x})$. The purpose of these is to approximate the unknown function $h(\acute{x})$:

$$h(\acute{x}) \approx \sum_{n=1}^{N} Q_n f_n(\acute{x}) \qquad (2.15)$$

for our specific case:

$$q_s(\acute{x}) \approx \sum_{n=1}^{N} Q_n f_n(\acute{x}) \qquad (2.16)$$

We substitute the right–hand side of (2.16) into (2.5):

$$\phi(x) = \int_{-a/2}^{a/2} \sum_{n=1}^{N} Q_n f_n(\acute{x}) \frac{1}{2\pi\epsilon_0} \ln\left(\frac{1}{\mid x - \acute{x} \mid}\right) d\acute{x} \qquad (2.17)$$

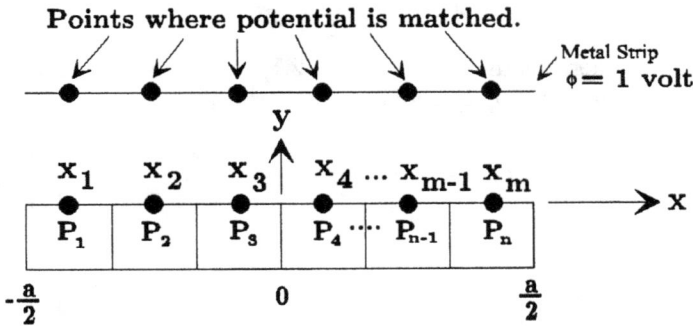

**Figure 2.2**   We divide the charged strip into $N$ regions of charge, also called *subdomains*. The use of pulse functions assumes the charge is constant over each subdomain. This produces one equation with $N$ unknowns. We obtain the same number of equations as unknowns by selecting $M$ observation points along the strip, which are all known to be at a potential of 1 volt. We will choose these points at the center of each pulse function.

The $Q_n$'s are constants representing values of charge/unit length, which are to be determined. We will approximate the shape of the charge/unit length distribution along $x$ (i.e., $f_n(\acute{x})$) using pulse functions, which are defined below:

$$P_n(\acute{z}) = \begin{cases} 1 & \text{if } \acute{z} \in (a_n, b_n) \\ 0 & \text{elsewhere} \end{cases} \tag{2.18}$$

Pulse functions cause the integrand to vanish anywhere outside of the interval $(a_n, b_n)$, allowing us to write (2.17) as:

$$\phi(x) = \int_{a_n}^{b_n} \sum_{n=1}^{N} \frac{Q_n}{2\pi\epsilon_0} \ln\left(\frac{1}{|x - \acute{x}|}\right) d\acute{x} \tag{2.19}$$

We know the potential of the strip is 1V everywhere on the strip. Therefore, $g(x)$ is a constant value of 1V.

$$\int_{a_n}^{b_n} \sum_{n=1}^{N} \frac{Q_n}{2\pi\epsilon_0} \ln\left(\frac{1}{|x - \acute{x}|}\right) d\acute{x} = 1 \tag{2.20}$$

The values of $Q_n$ are constants, which allows us to bring them out in front of the integral:

$$\sum_{n=1}^{N} \frac{Q_n}{2\pi\epsilon_0} \int_{a_n}^{b_n} \ln\left(\frac{1}{|x_m - \acute{x}|}\right) d\acute{x} = 1 \tag{2.21}$$

We can now distribute pulse functions along our charge–bearing strip. We divide the strip into $N$ pulses (Figure 2.2). When we use pulse functions, we essentially assume the charge is constant over each interval. This is sometimes referred to as a *stair step* approximation.

The $N$ pulse functions give us a single equation with $N$ unknown constants ($Q_n$'s). We can choose $M$ points along the strip, which are all known to be at the same potential of 1V. Because the chosen points are known to match the 1V value of the right–hand side of (2.21), this is known as *point matching*. We will choose the points (i.e., $x_m$'s)—where we observe the unity potential to be—at the center of each pulse function. By dividing the strip into $N$ pulses, we obtain subdomains of length $H$:

$$H = \frac{a}{N} \tag{2.22}$$

The integration limits are given as:

$$n = 1, 2, 3, \ldots, N \quad \begin{cases} a_n = -a/2 + (n-1)H \\ b_n = -a/2 + nH \end{cases} \tag{2.23}$$

The match points are given by:

$$x_m = -a/2 + (m - 1/2)H \text{ for } m = 1, 2, 3, \ldots, M \qquad (2.24)$$

We now have the same number of equations as unknowns if we choose $M = N$. These may be expressed in matrix form:

$$
\begin{vmatrix}
a_{11} & a_{12} & a_{13} & \cdots & a_{1N} \\
a_{21} & a_{22} & a_{23} & \vdots & a_{2N} \\
a_{31} & a_{32} & a_{33} & \vdots & a_{3N} \\
\vdots & \vdots & \vdots & \vdots & \vdots \\
a_{N,1} & a_{N,2} & a_{N,3} & \cdots & a_{N,N}
\end{vmatrix}
\begin{vmatrix}
Q_1 \\ Q_2 \\ Q_3 \\ \vdots \\ Q_{N-1} \\ Q_N
\end{vmatrix}
=
\begin{vmatrix}
b_1 \\ b_2 \\ b_3 \\ \vdots \\ b_{N-1} \\ b_N
\end{vmatrix}
\qquad (2.25)
$$

Where

$$a_{mn} = \frac{1}{2\pi\epsilon_0} \int_{a_n}^{b_n} \ln\left(\frac{1}{\mid x_m - \acute{x} \mid}\right) d\acute{x} \qquad (2.26)$$

$$b_m = 1 \qquad (2.27)$$

The system of equations in (2.25) was solved for the values of charge/unit length, using midpoint numerical integration. The results are presented in Figure 2.3 and plotted with the exact solution for comparison. We can see that the moment method solution is very accurate in the center and becomes less accurate as we approach the edge of the charged strip. The reader will note the singularity in the charge distribution at either edge of the strip. This type of singularity is common in electrostatic problems and has analogy in time–harmonic situations.

As some readers may have noted already, we do not have to integrate (2.26) numerically. We can integrate analytically and write $a_{mn}$ in closed form:

$$a_{mn} = \frac{K(b_n, x_m) - K(a_n, x_m)}{2\pi\epsilon_0} \qquad (2.28)$$

Where:

$$K(\alpha, \beta) = (\alpha - \beta)\left[1.0 - \ln \mid \alpha - \beta \mid\right] \qquad (2.29)$$

We might wonder if the closed–form solution (2.28) and (2.29) is more accurate than that of numerical integration. One way to compare the solutions is to calculate the total charge/unit length of our strip.

If we integrate the charge/unit length across the strip, we will obtain the total charge/unit length. This total charge/unit length can provide a meaningful quantity to assess the convergence of our moment method solution.

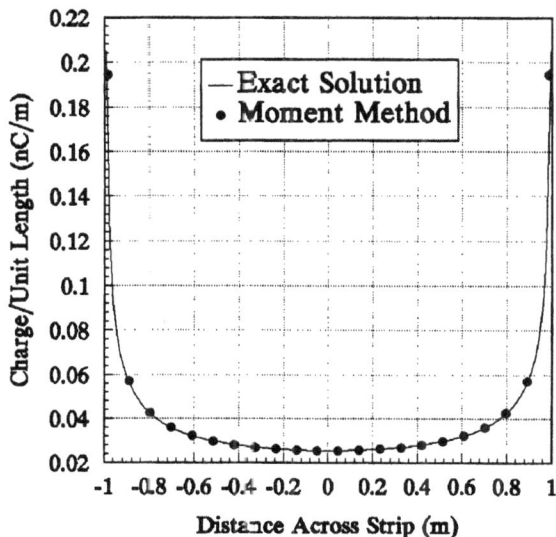

**Figure 2.3** A comparison of the values of charge/length obtained using the moment method and the exact solution. The moment method solution used 64 pulses and numerical integration.

In Table 2.1 we have the moment method solution for charge/unit length using midpoint numerical integration to evaluate (2.26) with five–place accuracy, and the closed–form solution (2.28) and (2.29).

We note that the total charge/unit length obtained is very close until we reach 64 segments. Above 64, the solutions are not quite as close. The extrapolated data shows a more prominent difference. The extrapolated values from the closed form are more stable above 64 segments than those of the numerical integration. The numerical integration leaves us with an uncertainty of 0.0010 pC/m. The best guess we might make is 80.2607 ±0.0005 pC/m from the numerical integration output. The closed–form output leads us to perhaps 80.2611 ±0.0001 pC/m.

Table 2.1

Moment Method Solution (Pulses with Point Matching)
for The Total Charge/Unit Length of a Charged Strip

| | (Numerical Integration) | | (Closed Form) | |
|---|---|---|---|---|
| M | Charge | Extrapolation | Charge | Extrapolation |
| | (pC/m) | (pC/m) | (pC/m) | (pC/m) |
| 2 | 64.0054 | 64.0054 | 64.0054 | 64.0054 |
| 4 | 71.2224 | 78.4395 | 71.2225 | 78.4395 |
| 8 | 75.4815 | 80.1742 | 75.4815 | 80.1741 |
| 16 | 77.8032 | 80.2642 | 77.8032 | 80.2643 |
| 32 | 79.0146 | 80.2605 | 79.0147 | 80.2609 |
| 64 | 79.6334 | 80.2612 | 79.6335 | 80.2611 |
| 128 | 79.9459 | 80.2602 | 79.9462 | 80.2610 |
| 256 | 80.1031 | 80.2611 | 80.1034 | 80.2612 |

## 2.2 GALERKIN'S METHOD

The estimates of total charge/unit length appear known within a very small uncertainty from the data in Table 2.1. When implementing the moment method,
the enforcement choice made was to match the value of the strip's potential at a
number of points along the strip. At points between there was no restriction on the
value of potential. There is no way to enforce the potential at every point along the
strip, but it is possible to enforce the boundary condition in an average fashion.

Instead of matching the value of potential at each point, we could choose to
integrate both sides of equation (2.20) over the domain of each pulse:

$$\int_{c_m}^{d_m} \int_{a_n}^{b_n} \sum_{n=1}^{N} \frac{Q_n}{2\pi\epsilon_0} \ln\left(\frac{1}{|x - \acute{x}|}\right) d\acute{x} \, dx = \int_{c_m}^{d_m} 1 \, dx \qquad (2.31)$$

We are essentially integrating the function enforcement over the subdomain
function, which contained only a single match point before. This is the same as
saying that all of the subdomain is equally important in terms of matching the
potential. We could choose a function to distribute the importance of potential
enforcement at each location along a subdomain, calling this function $W(x)$ for

*weighting function* and placing it into equation (2.31):

$$\int_{c_m}^{d_m} W(x) \int_{a_n}^{b_n} \sum_{n=1}^{N} \frac{Q_n}{2\pi\epsilon_0} \ln\left(\frac{1}{|x - \acute{x}|}\right) d\acute{x}\, dx =$$

$$\int_{c_m}^{d_m} W(x)\, 1\, dx \tag{2.32}$$

We can choose any function for $W(x)$. For example, we could choose a delta function centered at our match points: $\delta(x_m - x)$. Substituting this choice into (2.32) we obtain:

$$\int_{c_m}^{d_m} \delta(x_m - x) \int_{a_n}^{b_n} \sum_{n=1}^{N} \frac{Q_n}{2\pi\epsilon_0} \ln\left(\frac{1}{|x - \acute{x}|}\right) d\acute{x}\, dx =$$

$$\int_{c_m}^{d_m} \delta(x_m - x)\, 1\, dx \tag{2.33}$$

The sifting property of the delta function reduces the integral to the integrand evaluated at the point where the argument of the delta function becomes zero. This reduces (2.33) to:

$$\int_{a_n}^{b_n} \sum_{n=1}^{N} \frac{Q_n}{2\pi\epsilon_0} \ln\left(\frac{1}{|x_m - \acute{x}|}\right) d\acute{x} = 1 \tag{2.34}$$

We note that (2.34) is identical to equation (2.20). In other words, if we choose a delta function as a weighting function, the importance of satisfying the 1V potential is entirely concentrated at a single point $x_m$, which is identical to a point–matching solution. We can see that (2.33) is a more general expression of a moment method solution that contains point matching as a special case. Because $W(x)$ describes how we will enforce a known boundary condition, we sometimes refer to a weighting function as an *enforcement function* in this text.

We chose pulse functions to represent the shape of the charge over each subdomain. We can choose any function we wish for weighting. These are often chosen to facilitate closed–form integration. *Galerkin's method* is when the weighting functions are chosen as identical with the expansion functions. In our case, we chose pulse functions to expand the charge. We may now choose pulse functions as weighting functions in (2.32) and integrate. Carrying this out we obtain:

$$a_{mn} = \frac{K(c_m, b_n) - K(d_m, b_n) - K(c_m, a_n) + K(d_m, a_n)}{2\pi\epsilon_0} \tag{2.35}$$

Where:

$$K(\alpha, \beta) = \begin{cases} 0 & \text{if } (\alpha - \beta) = 0 \\ -(\alpha - \beta)^2 \left[\frac{\ln|\alpha - \beta|}{2} - \frac{3}{4}\right] & \text{elsewhere} \end{cases} \qquad (2.36)$$

$$b_m = H \qquad (2.37)$$

The new integration limits are given as:

$$m = 1, 2, 3, \ldots, N \quad \begin{cases} c_m = -a/2 + (m-1)H \\ d_m = -a/2 + mH \end{cases} \qquad (2.38)$$

**Table 2.2**

Moment Method Solution (Pulse Galerkin)
for The Total Charge/Unit Length of a Charged Strip

| M | Charge | Extrapolation |
|---|---|---|
|   | (pC/m) | (pC/m) |
| 2 | 68.9503 | 68.9503 |
| 4 | 74.2516 | 79.5528 |
| 8 | 77.9063 | 80.0704 |
| 16 | 78.6373 | 80.2480 |
| 32 | 79.4388 | 80.2600 |
| 64 | 79.8473 | 80.2610 |
| 128 | 80.0534 | 80.2607 |
| 256 | 80.1572 | 80.2617 |

The total charge/unit length obtained from the Galerkin method begins larger than the point–matching calculations (Table 2.2). The extrapolations once again refuse to converge in a monotonic fashion. The best estimate of charge/unit length we could obtain from the extrapolated data is $80.2612 \pm 0.0005$ pC/m. This is not as close an estimate as we obtained from point matching. The use of Galerkin's Method does not necessarily guarantee a better extrapolation.

## 2.3 SYMMETRY

In certain instances, we may know a problem has *symmetry*. For instance, we know *a priori* that the charge distribution is symmetric about the $x$-axis from previous calculations. From this we know that $q_s(\hat{x}) = q_s(-\hat{x})$. Recall equation (2.5):

$$\phi(x) = \int_{-a/2}^{a/2} \frac{q_s(\acute{x})}{2\pi\epsilon_0} \ln\left(\frac{1}{|x - \acute{x}|}\right) d\acute{x} \qquad (2.39)$$

The right–hand side may be rewritten as:

$$-\int_0^{-a/2} \frac{q_s(\acute{x})}{2\pi\epsilon_0} \ln\left(\frac{1}{|x - \acute{x}|}\right) d\acute{x} + \int_0^{a/2} \frac{q_s(\acute{x})}{2\pi\epsilon_0 V} \ln\left(\frac{1}{|x - \acute{x}|}\right) d\acute{x} \qquad (2.40)$$

We may change the sign of $\acute{x}$ in the left–hand term, which removes the negative sign out front and changes the upper limit to match the right–hand term:

$$\int_0^{a/2} \frac{q_s(-\acute{x})}{2\pi\epsilon_0 V} \ln\left(\frac{1}{|x + \acute{x}|}\right) d\acute{x} + \int_0^{a/2} \frac{q_s(\acute{x})}{2\pi\epsilon_0} \ln\left(\frac{1}{|x - \acute{x}|}\right) d\acute{x} \qquad (2.41)$$

Because $q_s(\acute{x}) = q_s(-\acute{x})$, we can now combine the two integrals:

$$\int_0^{a/2} \frac{q_s(\acute{x})}{2\pi\epsilon_0} \left[\ln\left(\frac{1}{|x + \acute{x}|}\right) + \ln\left(\frac{1}{|x - \acute{x}|}\right)\right] d\acute{x} \qquad (2.42)$$

The new integrand is known as a *symmetric kernel*. We can use this kernel only when we have symmetry in the problem. We might be curious if a symmetric kernel has a better convergence in this instance than that of previous methods. In Table 2.3 we see the results of implementing (2.28) and (2.29) using symmetry.

Table 2.3

Moment Method Solution
for The Total Charge/Unit Length of a Charged Strip
Pulse Point Match Closed Form Symmetric Kernel

| M | Charge (pC/m) | Extrapolation (pC/m) |
|---|---|---|
| 2 | 71.2225 | 71.2225 |
| 4 | 75.4815 | 79.7405 |
| 8 | 77.8032 | 80.2530 |
| 16 | 79.0147 | 80.2611 |
| 32 | 79.6335 | 80.2610 |
| 64 | 79.9462 | 80.2611 |
| 128 | 80.1033 | 80.2610 |
| 256 | 80.1821 | 80.2611 |

The convergence is monotonic with respect to the uncertainty obtained. We have a steady extrapolation of 80.26105 ±0.00005 pF. This is the most certain answer obtained so far. How close are we to the exact answer? We know the exact charge distribution from (2.11). Integrating, we obtain the total charge/unit length:

$$T_c = \frac{2\pi\epsilon_0}{\ln 2} \text{ pC/m} \tag{2.43}$$

Which to eight places beyond the decimal point is:

$$T_c = 80.26105482 \text{ pC/m}$$

We can see that the center value we obtain with the symmetric kernel is very close to a five–place accuracy beyond the decimal point. Of course, we would have no way of knowing this unless we have the exact answer with which to compare our moment method estimate. The problems of interest with which the moment method is implemented generally do not have an exact analytical solution. In Table 2.4, it is possible to compare the moment method estimates obtained using the methods presented for a charged strip. We can see that for this problem the symmetric kernel produced the best capacitance estimate. We note that the symmetric implementation allows us to sample the charged strip at double the rate we could without using symmetry.

## 2.4 CAPACITANCE OF A SQUARE CONDUCTING PLATE

As a final introductory example of moment method implementation, we will calculate the capacitance of a square conducting plate. We will analyze a plate of width and length $a$ (Figure 2.4) The plate resides in the $x$–$y$ plane at $z$=0 with the center of the plate coincident with the origin. The plate is assumed to have zero thickness. The electrostatic potential produced by the charge on the plate is given by:

$$\phi(x, y, z) = \int_{-a/2}^{a/2} \int_{-a/2}^{a/2} \frac{q_s(\acute{x}, \acute{y})}{4\pi\epsilon_0 R} \, d\acute{x} \, d\acute{y} \tag{2.44}$$

Where:

$q_s(\acute{x}, \acute{y})$ is the surface–charge density function (Coulombs/$m^2$)

$R = \sqrt{(x - \acute{x})^2 + (y - \acute{y})^2 + z^2}$

$\epsilon_0 = 8.854223 \cdot 10^{-12}$ Farads/meter

The difference in potential with respect to $\infty$ is the voltage. The potential on the plate is $\phi = V$. In our case we choose $V = 1$ V for convenience. With $z = 0$ we can write (2.44) as:

**Table 2.4**

Comparison of Total Charge/Unit Length Values Obtained
with Different Moment Method Implementations

| $T_c$ | Numercial Method |
|---|---|
| (pC/m) | (Integration – MM) |
| 80.2607 ±0.0005 | Numerical Integration Pulse Point Match |
| 80.2611 ±0.0001 | Closed Form Pulse Point Match |
| 80.2612 ±0.0005 | Closed Form Pulse Galerkin |
| 80.26105 ±0.00005 | Closed Form Pulse Point Match (Symmetric Kernel) |
| 80.26105482 | Exact Solution |

$$V = \int_{-a/2}^{a/2} \int_{-a/2}^{a/2} \frac{q_s(\acute{x}, \acute{y})}{4\pi\epsilon_0 \sqrt{(x - \acute{x})^2 + (y - \acute{y})^2}}\, d\acute{x}\, d\acute{y} \qquad (2.45)$$

We can divide the charged plate into subplates as shown in Figure 2.5. The number of subsections and match points is $N^2$. We will assume the charge is approximately constant over each sub-area. This allows us to define a two–dimensional pulse function:

$$P_n(\acute{x}, \acute{y}) = \begin{cases} 1 & \text{if } (\acute{x}, \acute{y}) \in S_n \\ 0 & \text{elsewhere} \end{cases} \qquad (2.46)$$

We chose the shape of the charge distribution over each subplate but not its magnitude. The two–dimensional pulse function is normalized. We will now introduce a set of constants $(Q_n)$ that define the magnitude of the charge over each subplate. This allows us to approximate $q_s(\acute{x}, \acute{y})$ as:

$$q_s(\acute{x}, \acute{y}) \approx \sum_{n=1}^{N^2} Q_n P_n(\acute{x}, \acute{y}) \qquad (2.47)$$

Where: $n = 1, 2, 3, \ldots, N^2$

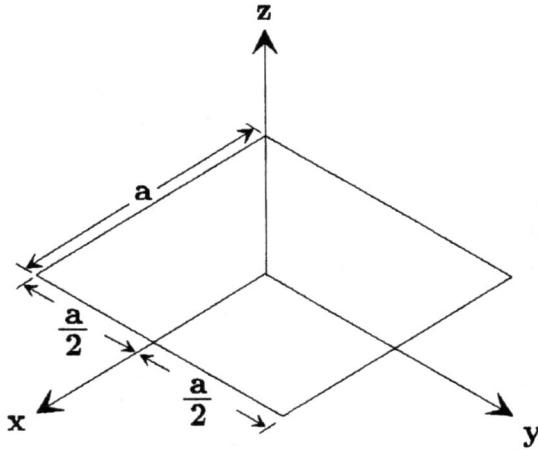

**Figure 2.4** A square conductive plate of width $a$ per side is raised to a constant potential $V$. The plate lies in the $x$–$y$ plane at $z = 0$.

placing (2.47) into (2.45):

$$1 = \sum_{n=1}^{N^2} \int_{-a/2}^{a/2} \int_{-a/2}^{a/2} \frac{Q_n P_n(\acute{x}, \acute{y})}{4\pi\epsilon_0 \sqrt{(x - \acute{x})^2 + (y - \acute{y})^2}} \, d\acute{x} \, d\acute{y} \qquad (2.48)$$

The range of the two–dimensional pulse functions reduce the integral to:

$$1 = \sum_{n=1}^{N^2} Q_n \int_{S_n} \int \frac{1}{4\pi\epsilon_0 \sqrt{(x - \acute{x})^2 + (y - \acute{y})^2}} \, dS_n \qquad (2.49)$$

We now have one equation in $N^2$ unknowns. As usual, we obtain $N^2$ equations by matching the potential at $N^2$ match points: $M_1$, $M_2$, $M_3$, ..., $M_{N^2}$, where: $(x_m, y_m) = M_m$.

We can identify our matrix elements:

$$a_{mn} = \int_{S_n} \int \frac{1}{4\pi\epsilon_0 \sqrt{(x_m - \acute{x})^2 + (y_m - \acute{y})^2}} \, dS_n \qquad (2.50)$$

and:

$$b_m = 1$$

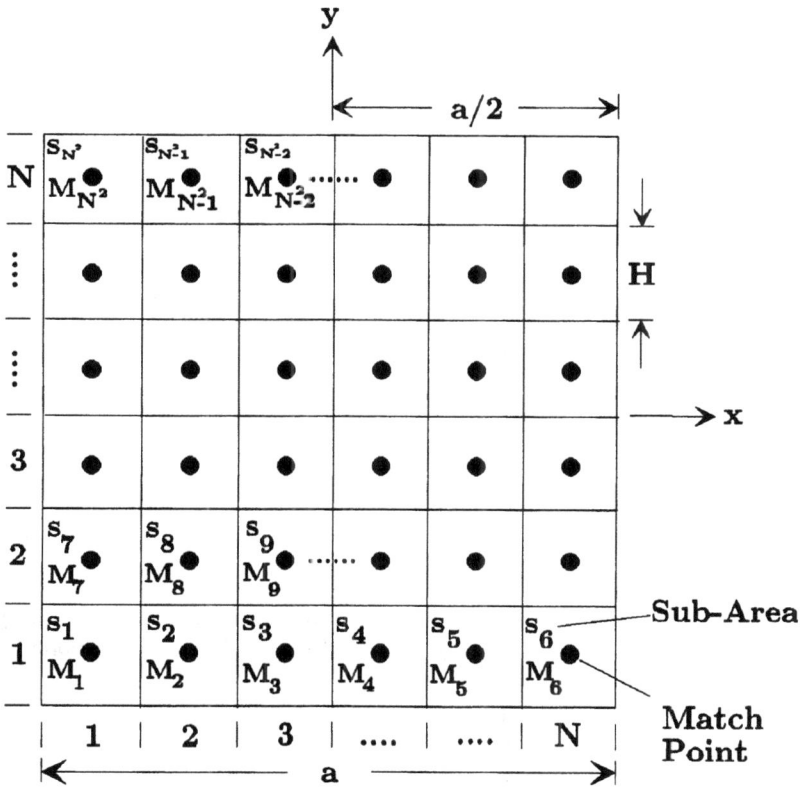

**Figure 2.5**  We can partition our conductive plate into $N^2$ subareas by dividing the $x$ and $y$ axis into $N$ subsections of length $H$. The subareas may be labeled $S_1, S_2, \ldots, S_{N^2}$. We may enforce the plate potential at $N^2$ match points, which lie in the center of each subarea.

Which produces this system of equations:

$$
\begin{vmatrix}
a_{11} & a_{12} & a_{13} & \cdots & a_{1N^2} \\
a_{21} & a_{22} & a_{23} & \vdots & a_{2N^2} \\
a_{31} & a_{32} & a_{33} & \vdots & a_{3N^2} \\
\vdots & \vdots & \vdots & \vdots & \vdots \\
a_{N^2,1} & a_{N^2,2} & a_{N^2,3} & \cdots & a_{N^2,N^2}
\end{vmatrix}
\begin{vmatrix}
Q_1 \\
Q_2 \\
Q_3 \\
\vdots \\
Q_{N^2-1} \\
Q_{N^2}
\end{vmatrix}
=
\begin{vmatrix}
b_1 \\
b_2 \\
b_3 \\
\vdots \\
b_{N^2-1} \\
b_{N^2}
\end{vmatrix}
\qquad (2.51)
$$

Table 2.5

Moment Method Solution
for The Capacitance of a Charged Conductive Plate

Capacitance

| M | Full Kernel | Extrapolation |
|---|---|---|
| | (pF) | (pF) |
| 2 | 70.3512 | 70.3512 |
| 4 | 75.4711 | 80.5911 |
| 8 | 78.3775 | 81.5148 |
| 16 | 79.9452 | 81.5999 |
| 32 | 80.7656 | 81.6143 |

The problem has symmetry about the $x$ and $y$ axes. We can use this symmetry to reformulate the problem with a symmetric kernel. When this is done we obtain:

$$a_{mn} = \int_{S_n} \int \frac{1}{4\pi\epsilon_0} K(x_m, \acute{x}, y_m, \acute{y})\, dS_n \qquad (2.52)$$

Where the symmetric kernel is:

$$K(x_m, \acute{x}, y_m, \acute{y}) =$$

$$\left[ \frac{1}{\sqrt{(x_m - \acute{x})^2 + (y_m - \acute{y})^2}} + \frac{1}{\sqrt{(x_m - \acute{x})^2 + (y_m + \acute{y})^2}} \right.$$

$$\left. + \frac{1}{\sqrt{(x_m + \acute{x})^2 + (y_m - \acute{y})^2}} + \frac{1}{\sqrt{(x_m + \acute{x})^2 + (y_m + \acute{y})^2}} \right] \qquad (2.53)$$

We integrate over 1/4 of the charged plate and multiply the total charge obtained using the moment method by 4 to calculate the capacitance of the entire plate. When the symmetric and nonsymmetric formulations are implemented using midpoint integration (with a plate length of 2 meters/side), we obtain the results in Tables 2.5 and 2.6.

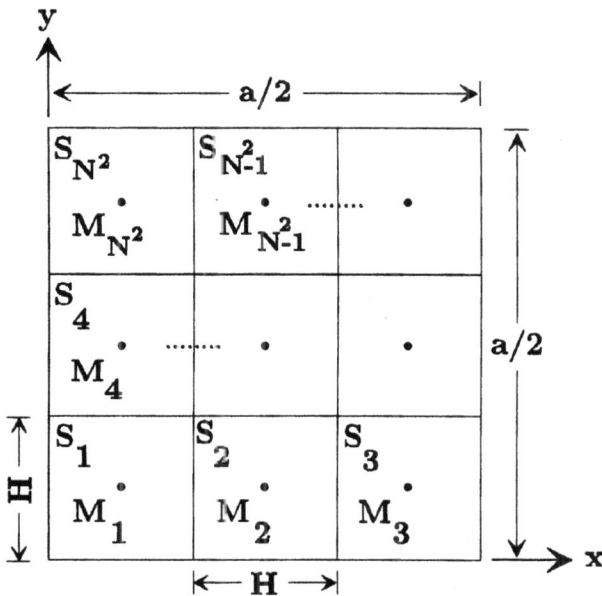

**Figure 2.6**   With the symmetric kernel we divide our conductive plate into $N^2$ subareas along the positive $x$ and $y$ axes. Each of the $N$ subsections is of length $H$. We only consider the upper one-quarter of the plate because of symmetry. The subareas may be labeled $S_1, S_2, .. , S_{N^2}$. We may enforce the potential of the plate at $N^2$ match points that lie in the center of each subarea.

We note that the sampling rate of the symmetric version is twice (per side) that of the full kernel. When $2M$ is 4, 8, 16, and 32 for the symmetric kernel, we note the results are essentially identical to those of the full kernel solutions at $M = 4, 8, 16,$ and 32. This is to be expected. Using the symmetric kernel we can calculate a 256 by 256 matrix and effectively perform a 1024 by 1024 full-kernel matrix calculation. This saves a considerable computer memory and produces a result that converges at a higher rate.

The extrapolated values of Table 2.5 converge but not perfectly monotonically. The difference between estimates decreases steadily. The best guess we can make is 81.61 ±0.01 pF. This is a bit conservative, but the best convergence we obtain is for two extrapolations.

Table 2.6

Moment Method Solution
for The Capacitance of a Charged Conductive Plate

| 2M | Symmetric Kernel | Extrapolation |
|----|------------------|---------------|
|    | (pF)             | (pF)          |
| 4  | 75.4710          | 75.4710       |
| 8  | 78.3775          | 81.2839       |
| 16 | 79.9452          | 81.5892       |
| 32 | 80.7656          | 81.6135       |
| 64 | 81.1877          | 81.6189       |

The symmetric kernel produces a result that leads us to think the answer might be closer to 81.62 pF. The $2M = 64$ calculation requires a 1024–by–1024 matrix, which is at the limit of many personal computers as this is written. The best estimate we can embrace is probably 81.62 ±0.1 pF.

Harrington has derived approximate $a_{mn}$ values for the full–kernel charged–plate problem:[2]

$$a_{nn} = \frac{H}{\pi\epsilon_0} \ln\left(1 + \sqrt{2}\right) \qquad (2.54)$$

$$a_{mn} \approx \frac{H^2}{4\pi\epsilon_0 \sqrt{(x_m - x_n)^2 + (y_m - y_n)^2}} \qquad (2.55)$$

The $x_n$ and $y_n$ are the center of the subarea being integrated (i.e., they are the central point of $S_n$). The reader who is interested in obtaining a feel for implementing the moment method should write a computer program using the closed–form approximations above. Note the differences among the full–kernel values of Table 2.5, the symmetric kernel of 2.6, and those obtained using (2.54) and (2.55).

Figure 2.7 is a three–dimensional plot of the charge on the plate produced with the full kernel (529 pulses). We can see the expected one over square root increase in the charge at each plate edge, with the largest charge magnitude being at each corner. The reader is encouraged to write computer programs of static problems that illustrate this further, such as wires or parallel strips.[3]

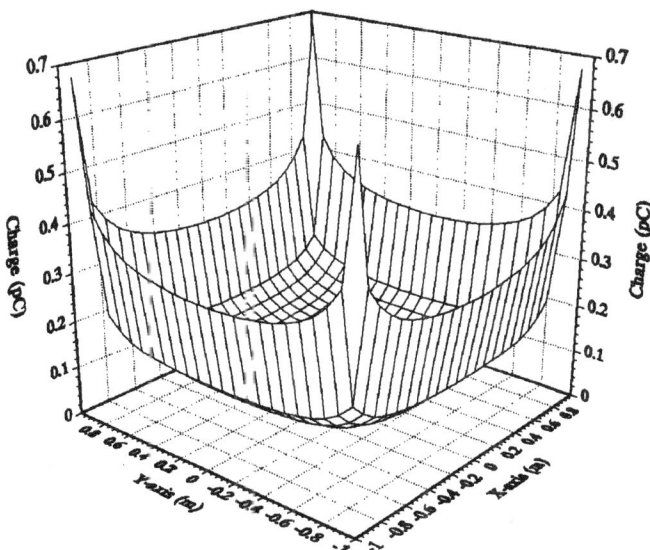

**Figure 2.7**  Three–dimensional plot of charge distribution on a 2–meter–square plate obtained with the moment method.

## 2.5  CONCLUDING REMARKS

We have implemented a number of moment method solutions in this chapter. The electrostatic problems lead us to a heuristic understanding of the moment method: many functions vary slowly enough that if we take a small–enough piece of them, they are essentially constant. If we have an equation of the form:

$$C = \int_a^b f(\acute{x}) K(x, \acute{x}) \, d\acute{x}$$

And we reduce the integration interval enough (i.e., $|a - b| \rightarrow 0$), then $f(\acute{x})$ becomes approximately constant and $K(x, \acute{x})d\acute{x}$ is known. This allows us to bring the constant $(f_c)$ out in front of the integral:

$$C = f_c \int_a^b K(x, \acute{x}) \, d\acute{x}$$

When we have reached this small of an interval, we may then use this interval size to create single equations with multiple unknown $f_c$'s:

$$C = f_{c_1} \int_{a_1}^{b_1} K(x, \acute{x})\, d\acute{x}$$

$$+ f_{c_2} \int_{a_2}^{b_2} K(x, \acute{x})\, d\acute{x}$$

$$+ f_{c_3} \int_{a_3}^{b_3} K(x, \acute{x})\, d\acute{x} \ldots$$

$$+ f_{c_N} \int_{a_N}^{b_N} K(x, \acute{x})\, d\acute{x}$$

This is similar to claiming:

$$\int_0^8 f(x)dx = \int_0^2 f(x)dx + \int_2^4 f(x)dx + \int_4^6 f(x)dx + \int_6^8 f(x)dx$$

The problem then allows us to obtain $N$ sets of equations by choosing $N$ points where we know the equation is satisfied. This crude explanation should help give the reader a feeling of what the moment method is about, and, I hope, help demystify the process from its name. In Chapter 3 we will become more elaborate in our moment method implementations. This heuristic explanation will help the reader realize the basis for what is to come.

# References

[1] Sadiku, Matthew N.O., *Numerical Techniques in Electromagnetics,* Boca Raton, Florida, CRC Press, 1992, pg. 315–334.

[2] Harrington, Roger F., *Field Computation by Moment Methods*, Malabar, Florida, Robert E. Krieger Publishing Company, 1968 (1983 Reprint Edition), pg. 27.

[3] Tsai, Leonard L. and Smith Charles E., "Moment Method in Electromagnetics for Undergraduates," *IEEE Transactions on Education*, Vol. E-21, No. 1, February 1978, pg. 14–22.

# Chapter 3

# Thin Wire Scattering

## 3.1 HALLEN'S EQUATION

The first electromagnetic scattering problem we will solve with the moment method is the determination of the radar cross section (RCS) of a wire in free space. In Figure 3.1 an electromagnetic wave traveling from the right encounters a wire at angle $\alpha$.

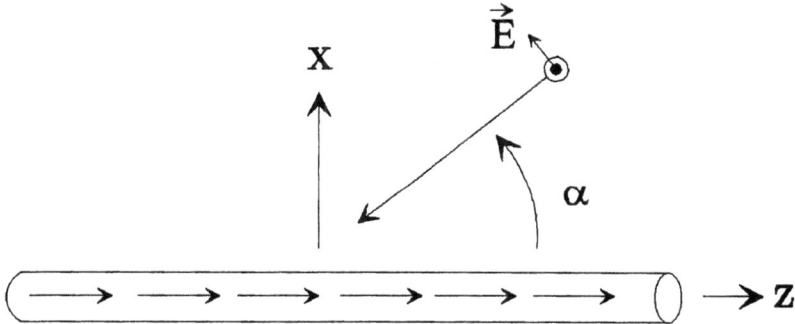

**Figure 3.1** An electromagnetic wave encounters a wire of radius $a$ and length $L$ at an angle $\alpha$. To maintain a zero tangential E–field along the conductor, currents are generated in the wire. The electric field produced by this current (i.e., $E^{scat}$) exactly cancels the incident electric field $E^{inc}$.

The tangential electric field at the surface of a perfect conductor must vanish (i.e., equal zero). At the surface of the wire the incident electromagnetic field ($E^{inc}$) must be canceled by a scattered electromagnetic field ($E^{scat}$). To meet this boundary condition, a current must flow on the surface of the wire to produce an opposing electric field. Therefore, the incident and scattered field exactly cancel at the surface of a perfect conductor. Expressed mathematically:

$$E^{inc} = -E^{scat} \qquad (3.1)$$

or

$$E^{inc} + E^{scat} = 0 \qquad (3.2)$$

We require an expression that relates the current induced on a wire by an incident electric field to the scattered field it produces. For a wire along the $z$-axis with a radius $a$ of length $L$, the relationship is:[1]

$$\frac{d^2 A(z)}{dz^2} + k^2 A(z) = j4\pi\omega\epsilon_0 E_z(z) \qquad (3.3)$$

$$A(z) = \int_{-L/2}^{L/2} I_z(\acute{z}) G(z, \acute{z}) \; d\acute{z} \qquad (3.4)$$

$$G(z, \acute{z}) = F(z - \acute{z}) \qquad (3.5)$$

$$F(R) = \int_0^{2\pi} \frac{e^{-jkR}}{R} d\acute{\phi} \qquad (3.6)$$

$$R = \sqrt{(z - \acute{z})^2 + \left(2a \sin \frac{\acute{\phi}}{2}\right)^2} \qquad (3.7)$$

The incident electric field along the conductor from a plane wave at angle $\alpha$ is:

$$E_z = e^{jkz \cos \alpha} \sin \alpha \qquad (3.8)$$

The solution of (3.3) consists of a complementary function $A_c$ and particular solution $A_p$.[2]

$$A_c = C_1 \cos kz + C_2 \sin kz \qquad (3.9)$$

$$A_p = A_1 \cos (kz \cos \alpha) + B_1 \sin (kz \cos \alpha) \qquad (3.10)$$

We may differentiate (3.10) twice, add the original equation multiplied by $k^2$, and equate it to the forcing function on the right-hand side of (3.3). This allows us to evaluate the constants $A_1$ and $B_1$. The final form of the particular solution is:

$$A_p(z) = j\frac{4\pi\omega\epsilon_0}{k^2 \sin\alpha} \cdot e^{jkz\cos\alpha} \tag{3.11}$$

The complete solution is:

$$A(z) = A_c(z) + A_p(z)$$

Which produces:

$$A(z) = C_1 \cos kz + C_2 \sin kz + j\frac{4\pi\omega\epsilon_0}{k^2 \sin\alpha} e^{jkz\cos\alpha} \tag{3.12}$$

We can equate this with (3.4) to produce a relationship between the current and the incident electric field:

$$\int_{-L/2}^{L/2} I_z(\acute{z})G(z,\acute{z}) \, d\acute{z} = C_1 \cos kz + C_2 \sin kz + j\frac{4\pi\omega\epsilon_0}{k^2 \sin\alpha} e^{jkz\cos\alpha} \tag{3.13}$$

This equation is a form of Hallen's equation. We will solve this equation for the special case of a plane wave at broadside ($\alpha = 90°$). This plane wave will drive the current in a symmetrical manner. This implies that $I(z) = I(-z)$, which means $A(z) = A(-z)$. For these conditions the constant $C_2$ vanishes.[3] [4] Hallen's equation for $\alpha = 90°$ with an assumed symmetric current is presented as:

$$\int_0^{L/2} I(\acute{z})G_s(z,\acute{z})d\acute{z} = j\frac{4\pi\omega\epsilon_0}{k^2} E^{scat}(z) + C_1 \cos(kz) \tag{3.14}$$

Where:

$$G_s(z,\acute{z}) = F(z-\acute{z}) + F(z+\acute{z}) \tag{3.15}$$

The function $G_s(z,\acute{z})$ is the symmetric Green's function. It is also referred to as the kernel of the equation. In many applications, the wire radius is very small compared with a wavelength. Equation (3.6) is often approximated using (3.16):

$$R \approx \sqrt{(z-\acute{z})^2 + a^2} \tag{3.16}$$

Using equation (3.15) with (3.4) and (3.5), produces what is known proverbially as the *reduced kernel.* Equation (3.17) is the symmetric form of the reduced kernel:

$$G_r(z,\acute{z}) = \frac{e^{-jk\sqrt{(z-\acute{z})^2+a^2}}}{\sqrt{(z-\acute{z})^2+a^2}} + \frac{e^{-jk\sqrt{(z+\acute{z})^2+a^2}}}{\sqrt{(z+\acute{z})^2+a^2}} \tag{3.17}$$

Note the reduced kernel is independent of $\acute{\phi}$. $I(\acute{z})$ is the current along the wire's length; $E^{scat}(z)$ is the electric field produced by this current distribution. $C_1$ is a constant that is determined by the boundary conditions of the problem. $R$ is the distance away from the wire at which we observe $E^{scat}$. The wire radius is $a$.

## 3.2 MOMENT METHOD SOLUTION (PULSE/DELTA)

Equation (3.14) is an integral equation that relates the wire current to the scattered E–Field. We can now implement a moment method solution. First, we expand the current using pulse functions:

$$I(\acute{z}) = \sum_{n=1}^{N} I_n P_n(\acute{z}) \tag{3.18}$$

We can place this into the left–hand side of equation (3.14). The integration is, as always, over the source (i.e., the current), which produces the E–field. We have assumed the current has a constant level over each subdivision of the wire. The pulse functions of (3.18) are defined as:

$$P_n(\acute{z}) = \begin{cases} 1 & \text{if } \acute{z} \in (a_n, b_n) \\ 0 & \text{elsewhere} \end{cases} \tag{3.19}$$

Substituting equation (3.18) into (3.14) produces:

$$\int_0^{\frac{L}{2}} \sum_{n=1}^{N} I_n P_n(\acute{z}) G_r(z, \acute{z}) d\acute{z} = j \frac{4\pi\omega\epsilon_0}{k^2} E^{scat}(z) + C_1 \cos(kz) \tag{3.20}$$

The pulse functions force the integrand to zero outside of each subdomain region $(a_n, b_n)$. The integral of a sum of functions is the sum of the integrals of each function separately. This allows us to interchange the integration and summation:

$$\sum_{n=1}^{N} \int_{a_n}^{b_n} I_n G_r(z, \acute{z}) d\acute{z} = j \frac{4\pi\omega\epsilon_0}{k^2} E^{scat}(z) + C_1 \cos(kz) \tag{3.21}$$

The pulse function determines the assumed shape of the current at each subdomain along the wire. The constant $I_n$ establishes the magnitude and phase. A constant can be brought out in front of the integral sign. We can also divide both sides by $\frac{1}{k^2} j 4\pi\omega\epsilon_0$ and rename the constant in front of the cosine function. $C_1$ now becomes $A$. The value of $A$ is not of importance to the calculation of the RCS.

$$\sum_{n=1}^{N} I_n \frac{k^2}{j4\pi\omega\epsilon_0} \int_{a_n}^{b_n} G_r(z, \acute{z}) d\acute{z} = E^{scat}(z) + A \cos(kz) \tag{3.22}$$

We can bring the cosine term to the left–hand side:

$$I_n \sum_{n=1}^{N} \frac{k^2}{j4\pi\omega\epsilon_0} \int_{a_n}^{b_n} G_r(z,\acute{z})d\acute{z} - A\cos(kz) = E^{scat}(z) \qquad (3.23)$$

We must now decide how to subdivide the current along the wire. The choice we will use is in Figure 3.2. The wire is broken into $N$ subdomains (segments). The beginning and ending subdomain length is $H/2$, and the intervening lengths are equal to $H$. Because we are using the symmetric Green's function, the half pulse at the origin is a full pulse when reflected about the $x$-axis. This also implies we have an incoming plane wave broadside to the wire ($\alpha = \frac{\pi}{2}$). The current is symmetric only when the magnitude and phase of an incoming wave is the same at all points along the wire. The relationship between $H$ and $N$ is given by:

$$H = \frac{L}{2N} \qquad (3.24)$$

The integration limits are given by:

$$n = 1 \quad \begin{cases} a_n = 0 \\ b_n = H/2 \end{cases} \qquad (3.25a)$$

$$1 < n < N \quad \begin{cases} a_n = H/2 + (n-2)H \\ b_n = H/2 + (n-1)H \end{cases} \qquad (3.25b)$$

$$n = N \quad \begin{cases} a_n = L/2 - H/2 \\ b_n = L/2 \end{cases} \qquad (3.25c)$$

We have now produced a single equation with $N+1$ unknowns; that is, we have $N$ $I_n$'s to determine along with the undetermined constant $A$ in front of the cosine term. The formulation of Hallen's integral equation implies that $A$ is determined using the boundary condition $I(L/2) = 0$. If we allow $I_n = 0$, then we will satisfy this condition.[5] This leaves us with one equation in $N$ unknowns. We may obtain the same number of equations as unknowns by using the known value of the electric field at $M$ places along the wire. At each point $z$ along the wire, the integrated current must produce the negative of the incident field to cancel it. When we pick a number of points to enforce this equality, we obtain the desired number of equations. We choose the field points along the wire to be at the center of each pulse function; the last pulse being the only exception. We still use a match point at the end of the wire. When we match the field of the equation at points along the wire, it is known as *point matching* or *co-location*. The match points $(z_m)$ are defined for each $m$ as:

$$m = 1 \quad z_m = 0 \qquad (3.26a)$$

$$1 < m < M \quad z_m = H/4 + (m-1)H \qquad (3.26b)$$

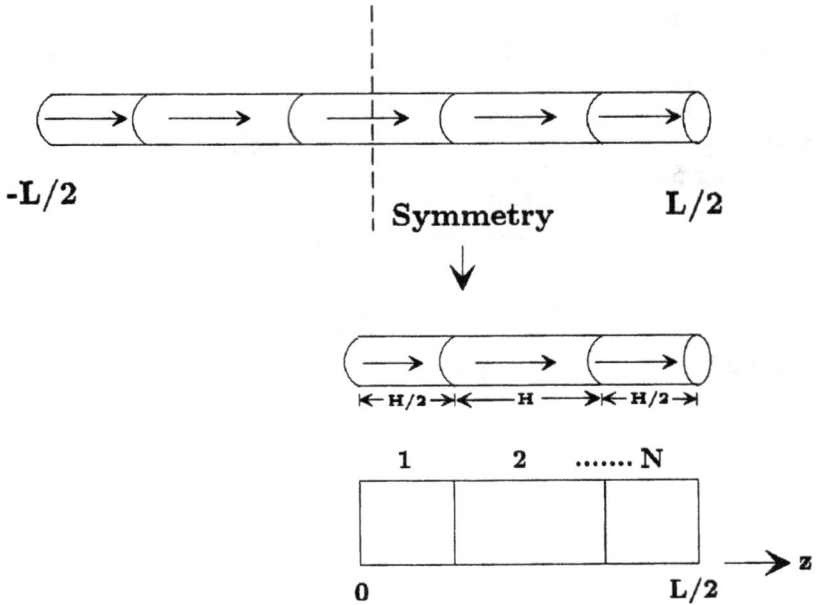

**Figure 3.2** The wire is divided into $N$ segments. The segment that originates at the center of the wire is of length $H/2$ as well as the terminating segment (i.e., the $N$th segment). The segments between are of length $H$. Because we use symmetry in this problem, the $H/2$ segment at the origin reflects into an $H$-sized pulse across the center of the wire.

$$m = M \quad z_m = L/2 \tag{3.26c}$$

We now have the same number of equations as unknowns. In matrix form:

$$
\begin{vmatrix}
a_{11} & a_{12} & \cdots & a_{1N} & c_1 \\
a_{21} & a_{22} & \vdots & a_{2N} & c_2 \\
a_{31} & a_{32} & \vdots & a_{3N} & c_3 \\
\vdots & \vdots & \vdots & \vdots & \vdots \\
a_{N,1} & a_{N,2} & \cdots & a_{N,N-1} & c_N
\end{vmatrix}
\begin{vmatrix}
I_1 \\
I_2 \\
\vdots \\
I_{N-1} \\
A
\end{vmatrix}
=
\begin{vmatrix}
b_1 \\
b_1 \\
\vdots \\
\vdots \\
b_M
\end{vmatrix}
\tag{3.27}
$$

Where:

$$a_{nm} = \frac{k^2}{j4\pi\omega\epsilon_0} \int_{a_n}^{b_n} G_r(z_m, \acute{z})d\acute{z} \tag{3.28}$$

And:

$$b_n = e^{jk\cos\alpha z_m} \tag{3.29}$$

We obtain $b_m$ by assuming a plane wave of unit amplitude:

$$E^{inc} = e^{jk\cos\alpha z} \tag{3.30}$$

Because we are using the symmetric kernel, we must have a symmetric incident field along the wire. The phase and magnitude of the incident field is constant along the wire when $\alpha = 90°$, which results in $b_m = 1$ for all values of $m$:

$$c_m = -\cos(kz_m) \tag{3.31}$$

We now have all the required information to solve for the unknown currents. Once we have the unknown currents, we can calculate the RCS of the wire. The definition of RCS for a three-dimensional case is:

$$\sigma = \lim_{R\to\infty} 4\pi R^2 \frac{|E^{scat}|^2}{|E^{inc}|^2} \tag{3.32}$$

We can calculate the RCS of the wire by first calculating the magnetic vector potential produced by the current:

$$A_z = \frac{e^{-jkR}}{4\pi R} \int_{-\frac{L}{2}}^{\frac{L}{2}} I(\acute{z})e^{-jk\acute{z}\cos\alpha}\sin\alpha \; d\acute{z} \tag{3.33}$$

Far from the source ($R \to \infty$) the E–field is:

$$E_z = -j\omega\mu A_z \tag{3.34}$$

Integrating over the current obtained from (3.27), using (3.33) and (3.34), allows us to calculate the scattered field ($E^{scat}$). We use symmetry to calculate the currents by assuming $I(\acute{z}) = I(-\acute{z})$. This causes (3.33) to become:

$$E^{scat} = -j\omega\mu\frac{e^{-jkR}}{4\pi R}2\int_0^{\frac{L}{2}} I(\acute{z})e^{-jk\acute{z}\cos\alpha} \; d\acute{z} \tag{3.35}$$

We can take the magnitude of the term out in front of the integration separately:

$$\left|E^{scat}\right|^2 = \left|\frac{\omega\mu}{4\pi R}\right|^2 \left|2\int_0^{\frac{L}{2}} I(\acute{z})e^{-jk\acute{z}\cos\alpha} \ d\acute{z}\right|^2 \tag{3.36}$$

The magnitude of the incoming plane wave is 1. By replacing $E^{inc}$ with 1 in equation (3.32) and substituting (3.36) into (3.32), we note that the $R$ dependence vanishes and we are left with:

$$\sigma = \frac{(\omega\mu)^2}{\pi}\left|\int_0^{\frac{L}{2}} I(\acute{z})e^{-jk\acute{z}\cos\alpha} \ d\acute{z}\right|^2 \tag{3.37}$$

We can numerically integrate using midpoint integration ($\alpha = \pi/2$) to obtain:

$$\sigma \approx \frac{(\omega\mu)^2}{\pi}\left|\frac{H}{2}I_1 + H\sum_{n=2}^{N-1} I_n\right|^2 \tag{3.38}$$

RCS ($\sigma$) is interpreted as an area. In general the units are square meters. $\sigma/\lambda^2$ is the normalized RCS. RCS is also described in logarithmic form as dB with respect to a 1 meter$^2$ reference. This quantity is called dBsm and is defined below:[6]

$$\sigma_{dBsm} = 10\log_{10}\sigma \tag{3.39}$$

### 3.2.1  Computational Results (Pulse/Delta)

Figures 3.3 and 3.4 present the current obtained for 64 and 128 segments respectively, using the reduced symmetric kernel (3.17). We note that the current for 64 segments is smooth except at the end of the wire where it deviates by hooking upward. The 128–segment solution has a ripple from the origin to the end. At the end of the wire we see a large oscillation appearing on the current.

The current obtained using the exact kernel (3.15) is presented in Figures 3.5 and 3.6. The current is quite smooth. The oscillation along the wire, which becomes maximum at the end for the reduced kernel solution, vanishes when the exact kernel is implemented. This leads us to suspect that this moment method solution is very sensitive to the approximation used to obtain the reduced kernel.

The computational results for the moment method solution using the reduced and exact kernel are presented in Table 3.1. The angle of the incoming plane wave is $\alpha = 90°$ and is polarized along the $z$–axis.

The very best answer we could hope to obtain from the reduced kernel data is from the unextrapolated numbers. Optimistically we could argue from the $M = 128$ data that $\sigma/\lambda^2 = 0.8041 \pm 0.0064$. The erratic behavior of the current for the 128–segment case has contributed to the instability of the RCS numbers for both the unextrapolated and extrapolated columns. This problem becomes more pronounced at $M = 256$ and 512.

The exact kernel solution has a much smoother current, and indeed the RCS calculation increases steadily as the number of segments is doubled. The increase does not extrapolate clearly until we reach our computational limit. The best estimate we can hope is $0.8438\pm0.0001$ for the normalized RCS of a $0.46\lambda$–long wire at broadside. The extrapolation is monotonic for just two computational doublings at m = 256 and 512. The apparent convergence of the unextrapolated calculations provides a certain measure of confidence that the m = 256 and 512 extrapolations provide an accurate RCS estimate.

Is there any other option we could exercise that might make the RCS increase more uniformly? We have the option of using whatever function we desire for the shape of our unknown current (i.e., the choice of basis function). We will next use triangular–shaped basis functions to try to obtain an extrapolation with certain monotonic convergence.

### Table 3.1

Hallen's Equation Pulse Expansion with Point Matching
LENGTH OF DIPOLE IN WAVELENGTHS:  0.46
RADIUS OF DIPOLE IN WAVELENGTHS:  0.005

| | REDUCED KERNEL | | EXACT KERNEL | |
|---|---|---|---|---|
| M | (RCS) | EXTRAPOLATION | (RCS) | EXTRAPOLATION |
| 4 | .0349 | .0349 | .0346 | .0346 |
| 8 | .2433 | .4517 | .2399 | .4452 |
| 16 | .6183 | 1.1739 | .6021 | 1.1373 |
| 32 | .8105 | .9820 | .7933 | .9705 |
| 64 | .8465 | .8081 | .8401 | .8257 |
| 128 | .8401 | .8137 | .8471 | .8245 |
| 256 | .8284 | .8099 | .8467 | .8439 |
| 512 | .8363 | .8655 | .8455 | .8438 |

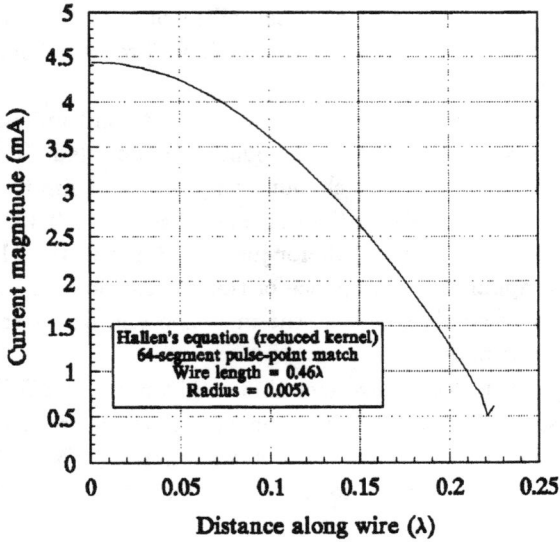

**Figure 3.3** Magnitude of the current obtained with 64 segments, using the reduced kernel, pulse–basis functions and point matching.

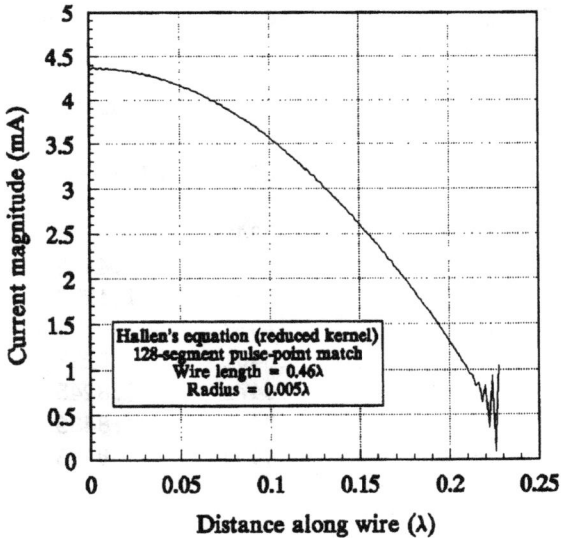

**Figure 3.4** Magnitude of the current obtained for a wire with 128 segments, using the reduced kernel, pulse–basis functions, and point matching.

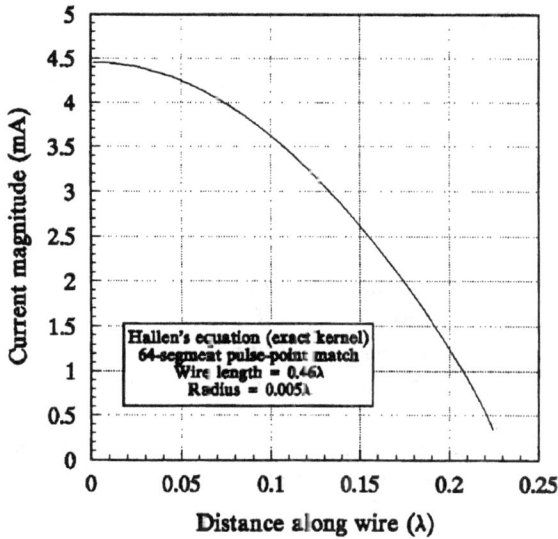

**Figure 3.5** Magnitude of the current obtained for a wire with 64 segments, using the exact kernel, pulse–basis functions. and point matching.

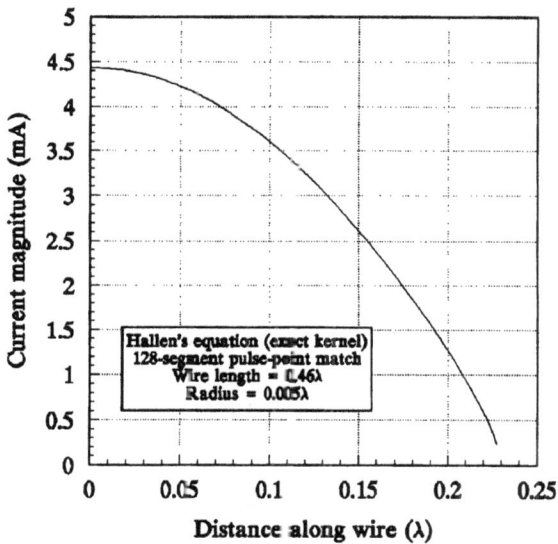

**Figure 3.6** Magnitude of the current obtained for a wire with 128 segments, using the exact kernel, pulse–basis functions, and point matching.

## 3.3 MOMENT METHOD SOLUTION (TRIANGLE/DELTA)

We can choose a different current shape over each subdivision. Triangles are often used—they provide a linear approximation of the current. The triangle functions are defined as:

$$T_n(\acute{z}) = \begin{cases} 1 - \frac{|\acute{z}-z_n|}{H} & \text{if } \acute{z} \in (a_n, b_n) \\ 0 & \text{elsewhere} \end{cases} \tag{3.40}$$

The implementation of triangular basis functions is illustrated in Figure 3.7. We can see they are overlapping triangles with a half triangle at the origin. From the symmetry of the problem, this produces a full triangle of current at the center, as we saw in the pulse point matching solution earlier. The match points are chosen at the top of each current triangle with one extra at the end of the wire. The triangle function itself enforces the boundary condition that the current must vanish at the end of the wire. This formulation has produced $N$ match points with $N-1$ triangles.

We still use equation (3.24) to calculate our $H$ value. The integration limits are:

$$n = 1 \quad \begin{cases} a_n = 0 \\ b_n = H \end{cases} \tag{3.41a}$$

$$n \neq 1 \quad \begin{cases} a_n = (n-2)H \\ b_n = a_n + 2H \end{cases} \tag{3.41b}$$

The match points are given by:

$$z_m = (m-1)H \tag{3.42}$$

Inserting the triangular functions into our previous solution process, we obtain the matrix elements:

$$a_{nm} = \frac{k^2}{j4\pi\omega\epsilon_0} \int_{a_n}^{b_n} T_n(\acute{z})G(z_m, \acute{z})d\acute{z} \tag{3.43}$$

And:

$$b_m = 1 \quad (\alpha = 90°) \tag{3.44}$$

And:

$$c_m = -\cos(kz_m) \tag{3.45}$$

We now have a defined system of equations that we can solve in the same manner as (3.27) which is repeated below:

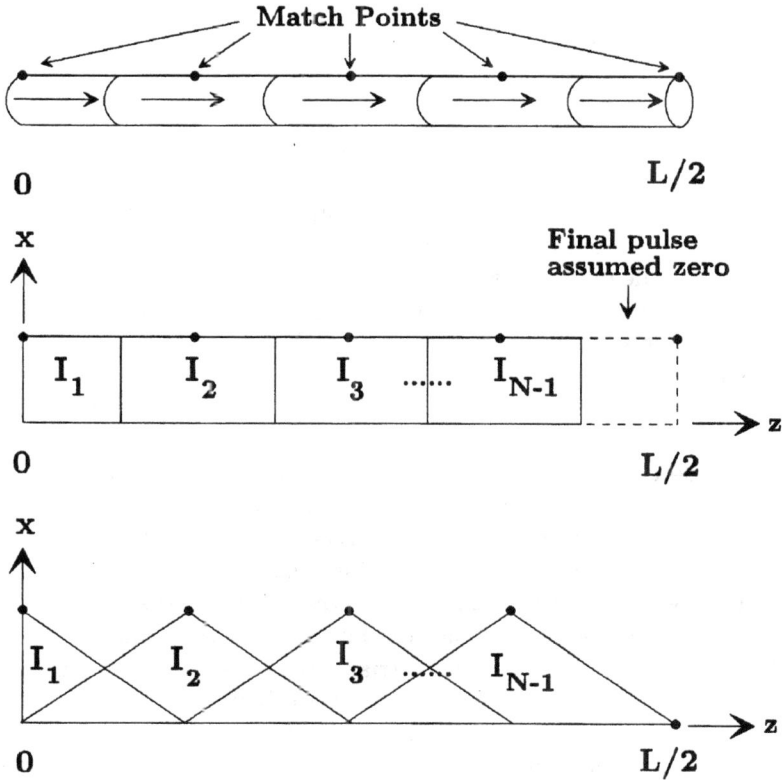

**Figure 3.7** Pulse expansion and triangle expansion. The wire is divided into $N$ segments. The segment that originates at the center of the wire is of length $H/2$. The segments are then of length $H$. Because we use symmetry in the problem, the $H/2$ segment at the origin reflects into an $H$-sized segment. A full pulse or triangle is across the center of the wire. Note we have $N$ match points with $N-1$ triangles, which satisfy the boundary condition of vanishing current at the end of the wire.

$$
\begin{vmatrix}
a_{11} & a_{12} & \cdots & a_{1N} & c_1 \\
a_{21} & a_{22} & \vdots & a_{2N} & c_2 \\
a_{31} & a_{32} & \vdots & a_{3N} & c_3 \\
\vdots & \vdots & \vdots & \vdots & \vdots \\
a_{N,1} & a_{N,2} & \cdots & a_{N,N-1} & c_n
\end{vmatrix}
\begin{vmatrix}
I_1 \\
I_2 \\
\vdots \\
\\
I_{N-1} \\
A
\end{vmatrix}
=
\begin{vmatrix}
1 \\
1 \\
1 \\
\vdots \\
\\
1
\end{vmatrix}
\qquad [3.27]
$$

## 3.3.1  Computational Results (Triangle/Delta)

Solution of this system of equations produces the current shown in Figures 3.8 and 3.9 for the reduced kernel. The 64–segment solution again shows a variation of the current for the reduced kernel as it did for the pulse point–match solution. The 128–segment solution has small oscillations along the current, which become large at the end of the wire. If we look carefully at Figure 3.4, we see similar oscillations at a much smaller amplitude.

The exact–kernel, triangle–point–match, moment–method current solutions in Figures 3.10 and 3.11 are quite smooth for both the M = 64 and 128 cases.

We can determine if the use of triangular basis functions has improved our RCS estimate by viewing the data contained in Table 3.2. From the reduced kernel data, we can see that the extrapolation again has not converged. The best guess we might obtain could be $\sigma/\lambda^2 = 0.845 \pm 0.018$. This is not as good a guess as the unreliable estimate we obtained using the reduced kernel with pulses and point matching. In both cases, the reduced kernel data has less stability than that of the exact kernel.

The exact kernel solution with triangles and point matching has produced smooth currents. We note a striking difference between the pulse point–matching solution in Tables 3.1 and 3.2. The raw data starts out at a much higher RCS and has a much smoother increase. We see this most prominently when we extrapolate the data. We can see from the extrapolated data in Table 3.2 that the extrapolated answer to four places is $\sigma/\lambda^2 = 0.8437$. The exact kernel solution using triangles and point matching becomes smooth enough after M = 32 to provide a monotonic extrapolation.

The reduced kernel for Hallen's equation has often been used to calculate antenna and scattering characteristics of wires. A mathematical proof has been offered to show that for thin wires, the two formulations are numerically indistinguishable. The numerical computations offered here indicate that this is not the case. Schelkunoff has demonstrated that an integral equation that uses the reduced kernel has no mathematical solutions.[7]  Hallen's equation is an ill-posed problem when the reduced kernel is used.[8] [9] These results suggest that the reduced kernel should not be used for thin wire scattering and antenna calculations.

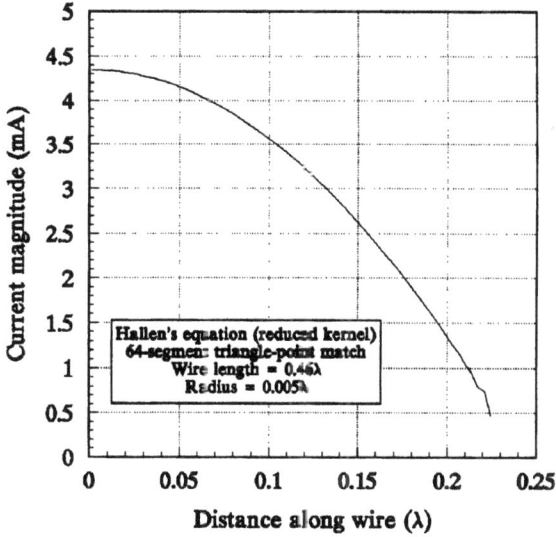

**Figure 3.8**  Magnitude of the current obtained for a wire with 64 segments, using the reduced kernel, triangle–basis functions, and point matching.

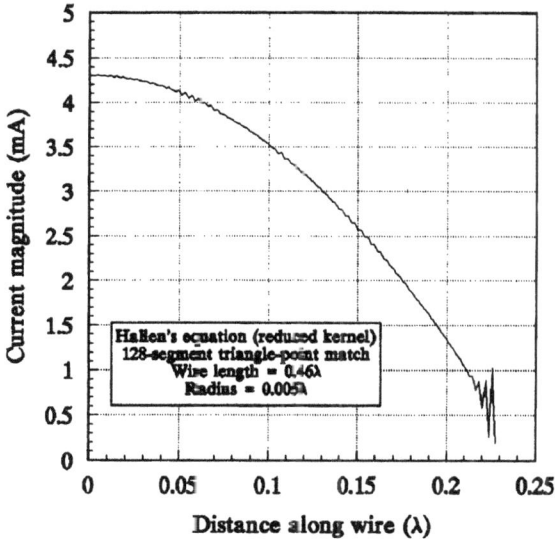

**Figure 3.9**  Magnitude of the current obtained for a wire with 128 segments, using the reduced kernel, triangle–basis functions, and point matching.

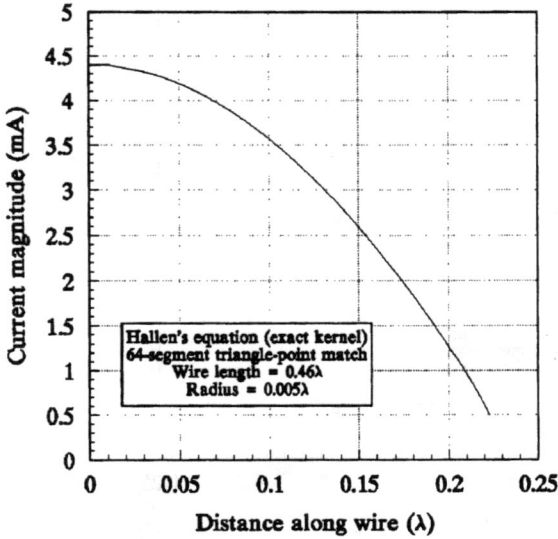

**Figure 3.10** Magnitude of the current obtained for a wire with 64 segments, using the exact kernel, triangle–basis functions, and point matching.

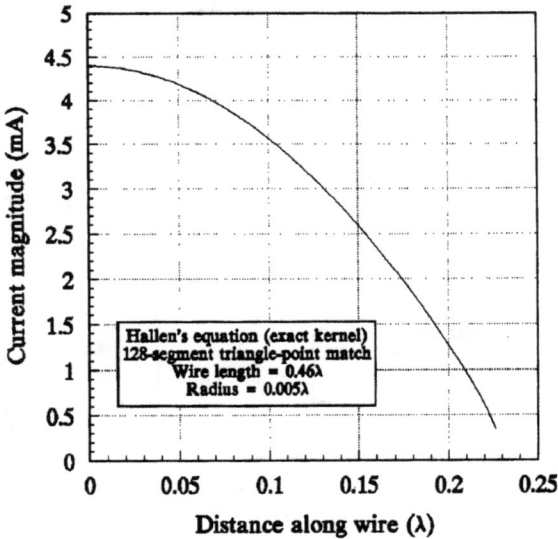

**Figure 3.11** Magnitude of the current obtained for a wire with 128 segments, using the exact kernel, triangle–basis functions, and point matching.

Table 3.2

Hallen's Equation Triangle Expansion and Point Matching

LENGTH OF DIPOLE IN WAVELENGTHS:  0.46

RADIUS OF DIPOLE IN WAVELENGTHS:  0.005

| | REDUCED KERNEL | | EXACT KERNEL | |
| --- | --- | --- | --- | --- |
| M | (RCS) | EXTRAPOLATION | (RCS) | EXTRAPOLATION |
| 2 | .8248 | .8248 | .8229 | .8229 |
| 4 | .8420 | .8592 | .8411 | .8592 |
| 8 | .8461 | .8472 | .8459 | .8480 |
| 16 | .8452 | .8416 | .8468 | .8466 |
| 32 | .8413 | .8337 | .8461 | .8440 |
| 64 | .8352 | .8239 | .8451 | .8437 |
| 128 | .8269 | .8125 | .8445 | .8437 |
| 256 | .8175 | .8017 | .8441 | .8438 |
| 512 | .8339 | .9028 | .8439 | .8437 |

For the case of scattering from a thin wire, the triangle point–match solution using the exact kernel is stable to cur computer's computational limit (512 x 512). The extrapolation values become monotonic at ≈ 0.8437 for the normalized RCS of the 0.46λ wire. We note that the reduced kernel solution diverges at 512 basis functions. The slight ripple seen in Figure 3.9 becomes a wild oscillation and the solution is essentially meaningless. We note that for m = 8, 16, and 32, the unextrapolated values are near the expected value and then diverge.

In Table 3.3, we have a summary of the solutions obtained for the scattering from a thin wire at broadside. The reduced kernel with pulses and point matching did not extrapolate with any degree of confidence. The exact kernel produced results that were of similar uncertainty but appeared to begin extrapolating. The use of triangles and point matching with the reduced kernel again produced unsatisfactory results. The exact kernel does extrapolate nicely.

The extrapolation numbers of the formulation using triangles and point matching may be used to calculate the RCS of a wire as a function of length. This data is presented in Figure 3.12. The length of maximum backscatter is approximately

**Table 3.3**

Comparison of RCS Values Obtained
with Different Moment Method Implementations

| RCS $(\sigma/\lambda^2)$ | Numerical Method (Kernel/Moment Method) |
|---|---|
| 0.8041 ±0.0064 (?)<br>No convergence | Reduced kernel<br>Pulse point match |
| 0.8438 ±0.0001 | Exact kernel<br>Pulse point match |
| 0.8450 ±0.018 (?)<br>No convergence | Reduced kernel<br>Triangle point match |
| 0.8437 (4 places) | Exact kernel<br>Triangle point match |

$L \approx 0.457\lambda$. We should also note that the scattering magnitude is sensitive to the value of radius used.

With all the above caveats, we must wonder about the fidelity this formulation has with nature. Measurements of scattering from thin wires at a number of lengths have been carried out by Chang and Liepa.[10] The frequency of measurement was 2.370 GHz. The radius to wavelength ratio is 0.00628 (i.e., $a/\lambda$ = 0.00628).

In Figure 3.13, we find a curve of the moment method RCS prediction for a wire as a function of wire length with experimental data points. The data is for the special case of $\alpha = 90°$. The moment method solution used triangular–basis functions, point matching, and the exact kernel. We see that the correlation between experiment and theory is less than a dB over almost the entire length examined. Despite the mathematical and practical uncertainties, the predictions look very good.

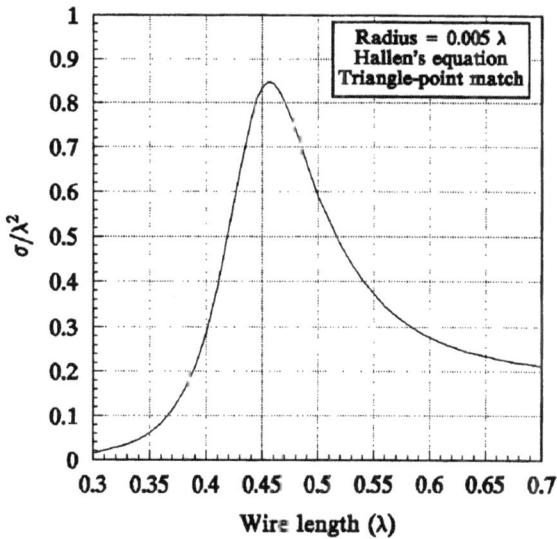

**Figure 3.12**  RCS (monostatic) of a wire as a function of the wire's length.  Triangular basis functions with point matching and Richardson's extrapolation.

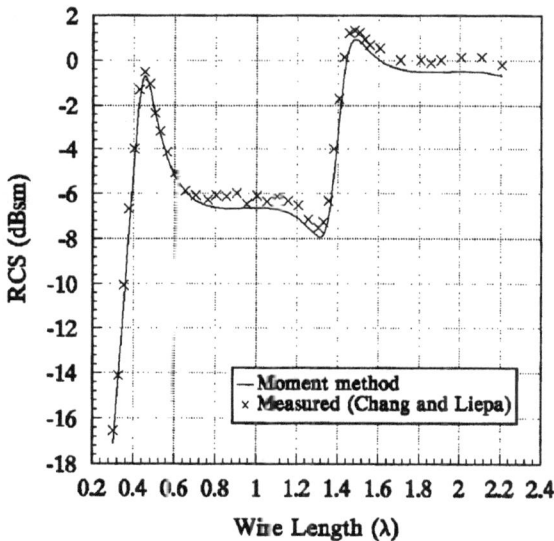

**Figure 3.13**  Calculated and measured RCS for a thin wire.  Hallen's equation triangles and point matching.

## 3.4 MOMENT METHOD SOLUTION AT ARBITRARY INCIDENCE

We now address the solution of Hallen's equation for an arbitrary incident angle. We will implement a triangle, point match solution of (3.13) repeated below:

$$\int_{-L/2}^{L/2} I_z(\acute{z})G(z,\acute{z}) \ d\acute{z} = C_1 \cos kz + C_2 \sin kz + j\frac{4\pi\omega\epsilon_0}{k^2 \sin\alpha}e^{jkz\cos\alpha} \qquad [3.13]$$

The implementation using pulse functions is left as an exercise. In Figure 3.14 is a current expansion scheme to use to solve (3.13) by the moment method. The number of current triangles is $N$. Each triangle is of length $2H$. The length of $H$ is:

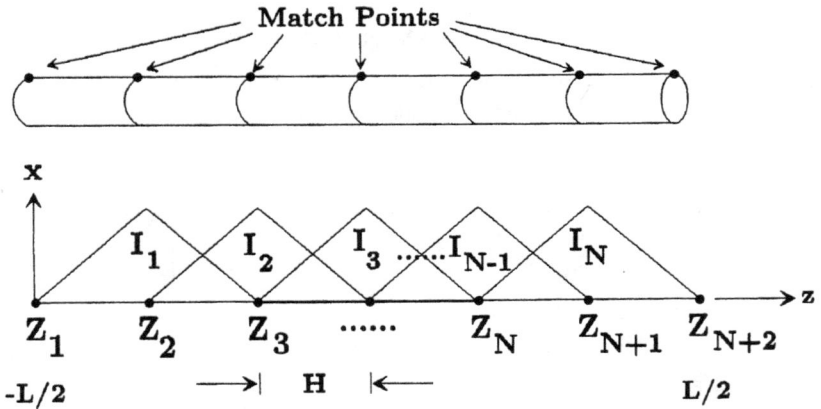

**Figure 3.14** Expansion using $N$ triangles with $N+2$ match points.

$$H = \frac{L}{(N+1)} \qquad (3.46)$$

We expand the current using triangles:

$$I(\acute{z}) = \sum_{n=1}^{N} I_n T_n(\acute{z}) \qquad (3.47)$$

Placing this into (3.13), we obtain:

$$\int_{-L/2}^{L/2} \sum_{n=1}^{N} I_n T_n(\acute{z}) G(z, \acute{z}) \ d\acute{z}$$

$$= C_1 \cos kz + C_2 \sin kz + j \frac{4\pi\omega\epsilon_0}{k^2 \sin \alpha} e^{jkz \cos \alpha} \tag{3.48}$$

The domain of the triangle functions is given by (3.40), repeated below:

$$T_n(\acute{z}) = \begin{cases} 1 - \frac{|\acute{z} - z_n|}{H} & \text{if } \acute{z} \in (a_n, b_n) \\ 0 & \text{elsewhere} \end{cases} \tag{3.40}$$

Which produces:

$$\sum_{n=1}^{N} I_n \int_{a_n}^{b_n} T_n(\acute{z}) G(z, \acute{z}) \ d\acute{z}$$

$$= C_1 \cos kz + C_2 \sin kz + j \frac{4\pi\omega\epsilon_0}{k^2 \sin \alpha} e^{jkz \cos \alpha} \tag{3.49}$$

The integration limits are:

$$a_n = -L/2 + (n-1)H \tag{3.50a}$$

$$b_n = a_n + 2H \tag{3.50b}$$

Where:

$$n = 1, 2, 3, \ldots N$$

In (3.49) we have one equation in $N + 2$ unknowns. The expansion scheme we have chosen allows $N + 2$ evenly spaced match points:

$$z_m = -L/2 + (m-1)H \tag{3.51}$$

Where:

$$m = 1, 2, 3, \ldots, N + 2$$

It also has the property that $I(\pm L/2) = 0$. This choice of match points produces $N + 2$ equations in $N + 2$ unknowns (3.52), which allows us to solve for the current:

$$\sum_{n=1}^{N} I_n \int_{a_n}^{b_n} T_n(\acute{z}) G(z_m, \acute{z}) \ d\acute{z} - C_1 \cos kz_m - C_2 \sin kz_m =$$

$$j\frac{4\pi\omega\epsilon_0}{k^2\sin\alpha}e^{jkz_m\cos\alpha} \tag{3.52}$$

In matrix form:

$$
\begin{vmatrix}
a_{11} & a_{12} & \cdots & a_{1N} & c_1 & d_1 \\
a_{21} & a_{22} & \vdots & a_{2N} & c_2 & d_2 \\
a_{31} & a_{32} & \vdots & a_{3N} & c_3 & d_3 \\
\vdots & \vdots & \vdots & \vdots & \vdots \\
a_{N+2,1} & a_{N+2,2} & \cdots & a_{N+2,N} & c_{N+2} & d_{N+2}
\end{vmatrix}
\begin{vmatrix}
I_1 \\
I_2 \\
\vdots \\
I_N \\
C_1 \\
C_2
\end{vmatrix}
=
\begin{vmatrix}
b_1 \\
b_1 \\
\vdots \\
\vdots \\
b_{N+2}
\end{vmatrix}
\tag{3.53}
$$

$$a_{m,n} = \int_{a_n}^{b_n} T_n(\acute{z})G(z_m,\acute{z})\ d\acute{z} \tag{3.54}$$

$$b_m = j\frac{4\pi\omega\epsilon_0}{k^2\sin\alpha}e^{jkz_m\cos\alpha} \tag{3.55}$$

$$c_m = -\cos kz_m \tag{3.56}$$

$$d_m = -\sin kz_m \tag{3.57}$$

We can use equation (3.33) to calculate RCS:

$$\sigma = \frac{(\omega\mu\sin\alpha)^2}{\pi}\left|\int_{-L/2}^{L/2}I(\acute{z})\sin\alpha e^{-jk\acute{z}\cos\alpha}d\acute{z}\right|^2 \tag{3.58}$$

in discreet form:

$$\sigma \approx \frac{(\omega\mu H\sin\alpha)^2}{\pi}\left|\sum_{n=1}^{N}I_n e^{-jkz_m\cos\alpha}\right|^2 \tag{3.59}$$

We can calculate the matrix elements using (3.54), (3.56), and (3.57) for any wire of a particular length and radius. This matrix may be augmented with (3.55) and the currents calculated for a given incident angle $\alpha$. Equation (3.59) provides an estimate of the RCS when observing the reflected wave at the same angle $\alpha$. Note that we need to calculate the [A] matrix only once. We can solve for the currents produced for any angle using (3.55) and [A]. When we step $\alpha$ through 360° we obtain the monostatic RCS for the wire.

In Figure 3.15, we have a plot of the RCS calculated using (3.53) and (3.59), along with experimental data taken at the University of Michigan by Chang and

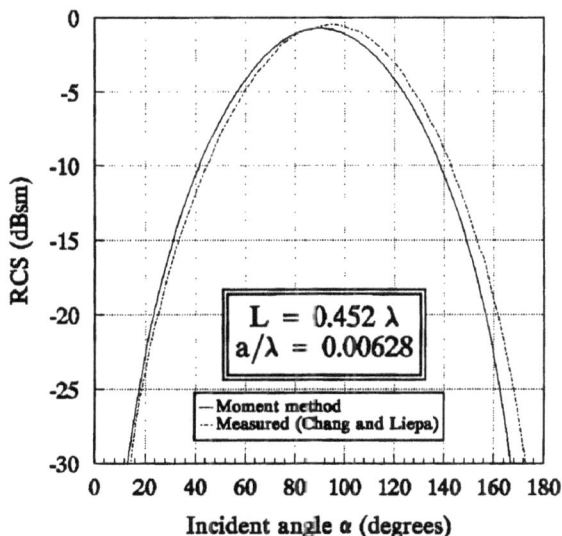

**Figure 3.15** RCS (monostatic) for a wire of length $0.452\lambda$, predicted by Hallen's equation, solved by the moment method with 126 triangular–basis functions and delta weighting. After: Chang and Liepa.

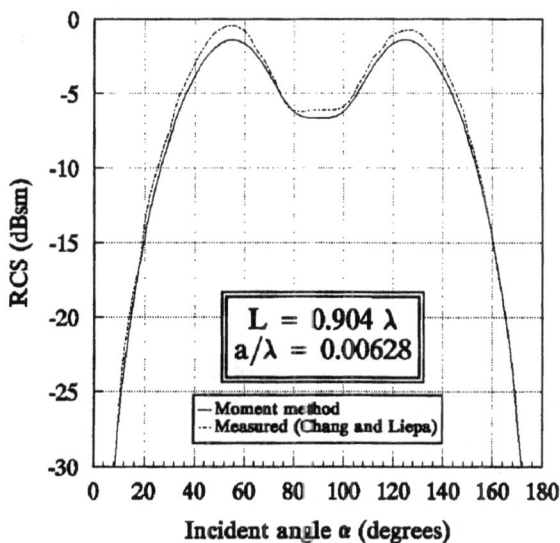

**Figure 3.16** RCS (monostatic) for a wire of length $0.904\lambda$, predicted by Hallen's equation, solved by the moment method with 126 triangular–basis functions and delta weighting. After: Chang and Liepa.

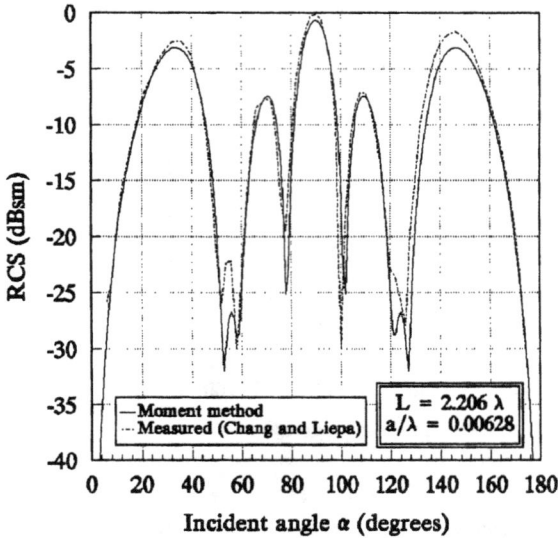

**Figure 3.17** RCS (monostatic) for a wire of length 2.206λ, predicted by Hallen's equation, solved
by the moment method with 126 triangular–basis functions and delta weighting.
After: Chang and Liepa.

Liepa. The number of basis functions used for the calculation is 126. No extrapolation was used. We note that the data shows great fidelity with measurement.
The difference in peak value between measured and calculated is less than 1 dB in
Figure 3.15. Figure 3.16, which presents a wire that is twice the length of that in
Figure 3.15, also shows about a one dB difference at the peaks. The final example is Figure 3.17. The moment method solution shows very good correlation with
the complex RCS pattern of the 2.206λ–long wire. These experimental correlations
should provide confidence in the usefulness of the moment method.

Hallen's equation is not the only formulation of a wire scatterer that allows a
moment method solution.[11]  Harrington presents a formulation that is generally
called a *mixed potential formulation*. This method writes the electric field in terms
of the magnetic vector potential and electric scalar potential. In this type of solution, we expand both current and charge along a wire to obtain a moment method
solution. The continuity equation allows us to write the electric scalar potential
in terms of the unknown current. We can factor out and solve the unknown current with the moment method. This solution method will be introduced when we
analyze the scattering from a transverse electric polarized wave in the next chapter.

# References

[1] Elliot, Robert S., *Antenna Theory and Design,* Englewood Cliffs, New Jersey, Prentice Hall, 1981, pg. 284–286.

[2] Zill, Dennis G., *A First Course in Differential Equations with Applications,* 3rd Edition, PWS–Kent Publishing Co., 1986, pg. 209.

[3] Booton, Richard C., Jr., *Computational Methods for Electromagnetics and Microwaves,* 1st edition, New York, John Wiley & Sons. 1992, pg. 135.

[4] Kong, Jin Au, *Electromagnetic Wave Theory,* John Wiley & Sons, 1986, pg. 470–473.

[5] Elliot, Robert S., *Antenna Theory and Design,* Englewood Cliffs, New Jersey, Prentice Hall, 1981, pg. 289,

[6] Knott, Shaeffer and Tuley, *Radar Cross Section,,* Norwood, MA, Artech House, 1985, pg. 50.

[7] Schelkunoff, Sergei A., *Advanced Antenna Theory,* New York, John Wiley & Sons, 1952, pg. 149–150.

[8] Booton, Richard C., Jr. "The Ill-Posed Nature of Hallen's Equation," *URSI Boulder Abstracts* 1993.

[9] Booton, Richard C., Jr. "More on the Ill-Posed Nature of Hallen's Equation," *URSI Boulder Abstracts,* 1994.

[10] Chang S., and Leipa, V. V. *Measured Backscattering Cross Section of Thin Wires,* University of Michigan Technical Report, 8077-4-T, May 1967.

[11] Harrington, Roger F., *Field Computation by Moment Methods,* Malabar, Florida, Robert E. Krieger Publishing Company, 1968 (1983 Reprint Edition), Chapter 4.

# Chapter 4

# Scattering From Conductive Strips

## 4.1 RCS OF PERFECTLY CONDUCTIVE STRIP

### 4.1.1 TM Polarization

In Chapter 2, we explored the charge distribution produced when a strip of width $a$ has a surface charge that creates a known potential $\phi$ on the strip. The electrostatic solution provided insight into how charge distributes itself in general. We saw that charge builds up at corners and edges.

We next calculate the current produced on the same strip by an incoming electromagnetic plane wave.[1] The currents generated on an infinite strip for the time–harmonic case will provide insight into how current is distributed along the edges and corners of a conductive sheet.

The same perfectly conducting strip in Figure 2.1 is found in Figure 4.1 on the following page.

The incoming electromagnetic wave is polarized with an electric field parallel to the $z$–axis. This polarization therefore produces a current on the strip that flows along the $Z$–axis. The magnetic field of this wave is entirely in the $x$–$y$ plane, and is therefore transverse to the $z$ axis. We refer to this as a transverse magnetic (TM) polarized wave. Other authors often refer to this as an $E$–wave or an $E$–polarized wave because the electric field is along the strip (i.e., parallel to it).

The magnetic vector potential of a current flowing along the strip is given by:

$$A_z = \frac{\mu}{4j} \int_{-a/2}^{a/2} I_z(\acute{x}) H_0^2(k \mid x - \acute{x} \mid) \, d\acute{x} \tag{4.1}$$

Where:

$K = \frac{2\pi}{\lambda}$ free-space wave number.

$G(x, \acute{x}) = \frac{1}{4j} H_0^2(k \mid x - \acute{x} \mid)$ 2D free–space Green's function.

$H_0^2(x)$ is a Hankel function of the second kind 0th order

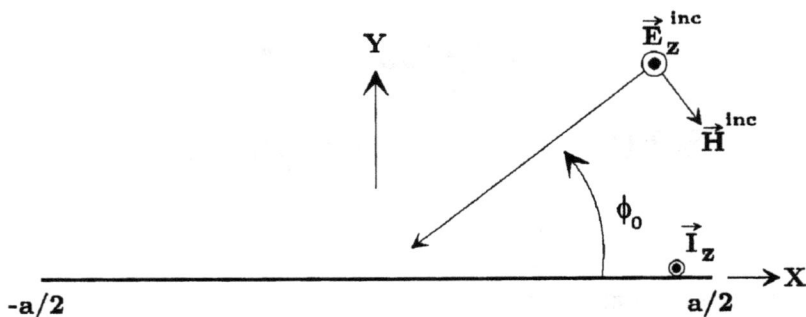

**Figure 4.1** A metal strip of width $a$ is encountered by an incoming plane wave that has a polarization with its magnetic field transverse to the $z$-axis. We refer to this as a TM polarized wave. The strip is assumed to be of infinite length in the $\pm z$ direction. We are interested in determining the current distribution across the strip. The induced current flows in the $z$ direction.

In Chapter 3, we were introduced to a three–dimensional free–space Green's function (3.5). An analogous free–space Green's function exists for two–dimensional problems.[2] [3] A Hankle function of the first kind, $H_0^1(x)$, represents an inward traveling–wave in two dimensions, and is analogous to $e^{jkR}$. A Hankel function of the second kind, $H_0^2(x)$, represents an outward–traveling wave and is analogous to $e^{-jkR}$.

The electric field is given by:

$$E_z(x) = j\omega A_z(x) \tag{4.2}$$

or

$$E(x) = \frac{\omega\mu}{4} \int_{-a/2}^{a/2} I_z(\acute{x}) H_0^2(k \mid x - \acute{x} \mid) \, d\acute{x}$$

We expand the current in the usual moment method manner (Figure 4.2) using pulse functions. In this case:

$$I_z(\acute{x}) = \sum_{n=1}^{N} I_n P_n \tag{4.3}$$

Inserting (4.3) into (4.1) and using (4.2) and (3.1) we obtain:

$$-E_z^{inc}(x) = \sum_{n=1}^{N} I_n \frac{\omega\mu}{4} \int_{a_n}^{b_n} H_0^2(k \mid x - \acute{x} \mid) \, d\acute{x} \qquad (4.4)$$

We divide the strip into $N$ segments, each of length $H$:

$$H = \frac{a}{N}$$

The integration limits are given by:

$$a_n = -a/2 + (n-1)H \qquad (4.5)$$

$$b_n = -a/2 + nH \qquad (4.6)$$

For $n = 1, 2, 3, \ldots, N$.

We enforce the electric field at $N$ match points to obtain the same number of equations as unknowns. The match points are given as:

$$x_m = -a/2 + \left(m - \frac{1}{2}\right)H \qquad (4.7)$$

for $m = 1, 2, 3, \ldots, N$

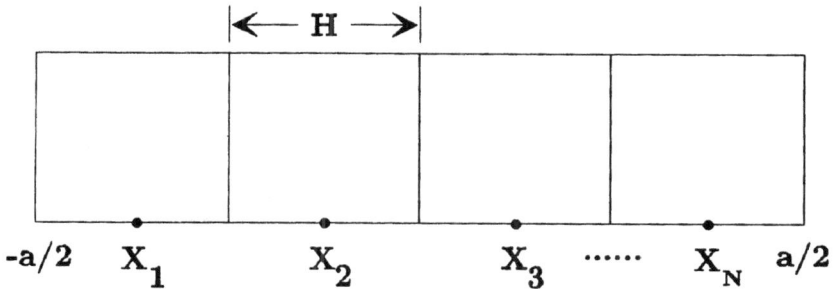

**Figure 4.2** The metal strip is divided into $N$ segments of length $H$ over which the current is assumed to be constant. At the center of each pulse, we match the field value to obtain an equal number of equations and unknowns.

In (4.4), we can now readily identify the matrix elements for the TM case as:

$$a_{mn} = \frac{\omega\mu}{4} \int_{a_n}^{b_n} H_0^2(k \mid x_m - \acute{x} \mid) \, d\acute{x} \qquad (4.8a)$$

and

$$b_m = -e^{jkx_m \cos\theta_0} \qquad (4.8b)$$

### 4.1.2 TE Polarization

The second case of interest is that of a plane wave polarized with its electric field transverse to the $z$ axis. This is illustrated in Figure 4.3. The electric field lies entirely in the $x$–$y$ plane and is therefore transverse to the $z$ axis. We refer to this as a transverse electric (TE) polarized wave. Other authors sometimes call this an $H$–wave or $H$–polarized wave, because the magnetic field of the wave is along the strip.

The equation that relates the scattered field to the induced current is:

$$-\vec{E}^{inc}(\vec{r}) \bullet \hat{x} = \left[ -j\omega\vec{A}(\vec{r}) - \vec{\nabla}\phi(\vec{r}) \right] \bullet \hat{x} \qquad (4.9)$$

Where for the case of our strip: $\vec{r} = x\hat{x}$

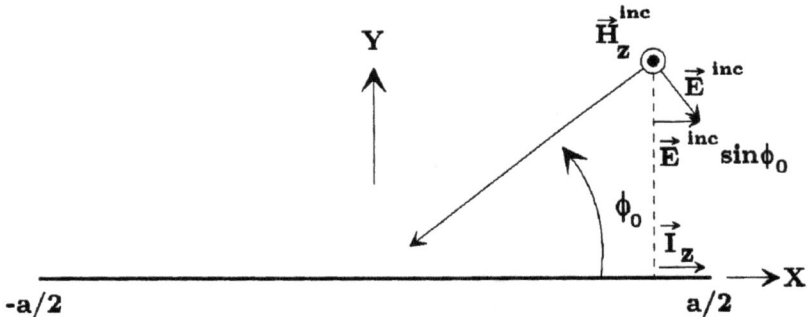

**Figure 4.3** A metal strip of width $a$ is encountered by an incoming EM plane wave that has its electric field transverse to the $z$ axis. We refer to this as a TE polarized wave. The strip is assumed to be of infinite length in the $\pm z$ direction. We are interested in determining the current distribution across the strip. The induced current flows in the $x$ direction.

The magnetic vector potential and electric scalar potential is given by (4.10) and (4.11), respectively:

$$\vec{A}(\vec{r}) = \frac{\mu}{4j} \int_{-a/2}^{a/2} \vec{I}(\vec{r})H_0^{(2)}(k \mid \vec{r} - \vec{r} \mid) \, d\acute{x} \qquad (4.10)$$

$$\phi(\vec{r}) = \frac{1}{4j\epsilon} \int_{-a/2}^{a/2} \rho(\vec{r})H_0^{(2)}(k \mid \vec{r} - \vec{r} \mid) \, d\acute{x} \qquad (4.11)$$

The current density $\vec{I}(\vec{r}) = I\hat{x}$ is related to the surface–charge density through the equation of continuity:

$$\rho(\vec{r}) = \frac{j}{\omega}\frac{d\vec{I}(\vec{r})}{d\hat{x}} \qquad (4.12)$$

The TM case requires that we consider only a magnetic vector potential term. The TE case has both vector and scalar potential terms. Using (4.9), (4.10), (4.11), and (4.12), we can write the following integro–differential equation:

$$\underbrace{-\frac{\omega\mu}{4}\int_{-a/2}^{a/2}\vec{I}(\acute{x})H_0^{(2)}(k\mid\vec{r}-\vec{r}\mid)\,d\acute{x}}_{\text{Magnetic vector potential}}$$

$$\underbrace{-\frac{d}{dx}\frac{1}{4j\epsilon}\int_{-a/2}^{a/2}\left[\frac{j}{\omega}\frac{d\vec{I}(\acute{x})}{d\acute{x}}\right]H_0^{(2)}(k\mid\vec{r}-\vec{r}\mid)\,d\acute{x}}_{\text{Electric scalar potential}}$$

$$= -\vec{E}^{inc}(\vec{r}) \qquad (4.13)$$

The expansion of current is written in the usual manner (using pulse expansion functions):

$$I(\vec{r}) = \sum_{n=1}^{N} I_n P_n(\vec{r}) \qquad (4.14)$$

We expand the charge and current in Figure 4.4. The integrand of the second term in (4.13) contains the current representation of charge by using the continuity equation (4.12). We can approximate the derivative of the current with a finite difference.

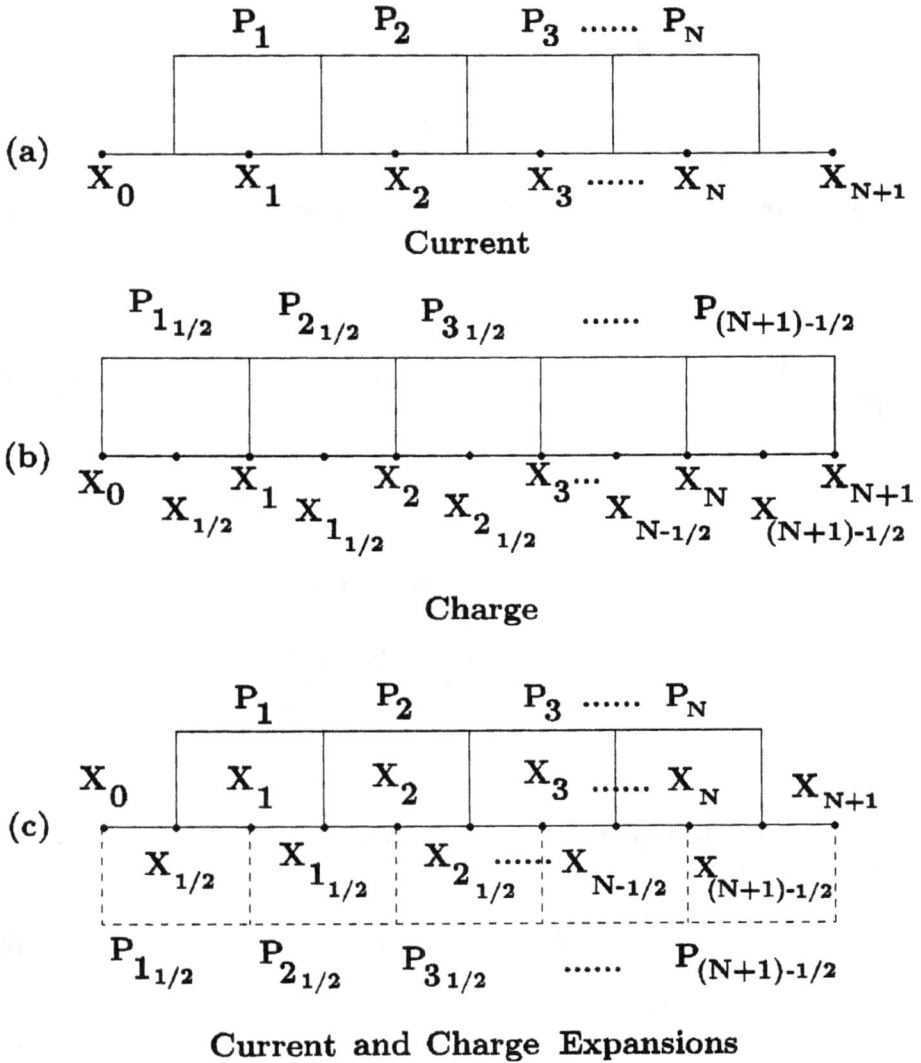

**Figure 4.4** (a) The current is represented using $N$ pulses and $N+1$ match points. The current pulses are labeled $P_1 \rightarrow P_N$. (b) The charge is expanded into $N+1$ pulses. These are labeled $P_{1/2} \rightarrow P_{(N+1)-1/2}$. (c) The two expansion schemes are combined with the match points into a single figure.

This allows us to express the charge function in terms of the unknown current constants.

$$\rho(\vec{r}) = \frac{j}{\omega} \sum_{n=1}^{N+1} \frac{I_n - I_{n-1}}{\Delta x_{n-1/2}} P_{n-1/2}(\vec{r})$$

(4.15)

Where: $\Delta x_{n-1/2} = x_n - x_{n-1}$

And

$P_{n-1/2}(x)$ extends from $x_{n-1}$ to $x_n$ with a value of unity.

The current at $x_0$ and $x_{N+1}$ must vanish (i.e., $I_0$ and $I_{N+1} = 0$). This boundary condition for the current, and an examination of the domain of the charge pulse functions, allows us to write:

$$\rho(\vec{r}) = \frac{j}{\omega} \sum_{n=1}^{N} I_n \left[ \frac{P_{n-1/2}(\vec{r})}{\Delta x_{n-1/2}} - \frac{P_{n+1/2}(\vec{r})}{\Delta x_{n+1/2}} \right]$$

(4.16)

We can substitute (4.16) into the second term of (4.13). This removes the explicit derivative inside the integral; but the second term of equation (4.13) still contains a derivative. We note that the derivative operates upon an integral that represents the scalar potential $\phi$. We can choose any weighting function we want. If we use pulse functions, an integral of this form is created:

$$\int_{x_{m-1/2}}^{x_{m+1/2}} \frac{d\phi}{dx} dx = \phi(x_{m+1/2}) - \phi(x_{m-1/2})$$

(4.17)

Thus, the derivative acting upon the potential in equation (4.13) disappears, and the scalar potential breaks out into two terms:

$$\underbrace{-\frac{\omega\mu}{4} \int_{-a/2}^{a/2} I(\acute{x}) H_0^{(2)}(k \mid \vec{r} - \acute{r} \mid) d\acute{x}}_{A}$$

$$\underbrace{\frac{1}{4j\epsilon} \int_{-a/2}^{a/2} \left[ \frac{j}{\omega} \frac{dI(\acute{x})}{d\acute{x}} \right] H_0^{(2)}(k \mid x_{m+1/2} - \acute{x} \mid) d\acute{x}}_{\phi(x_{m+1/2})}$$

$$\underbrace{-\frac{1}{4j\epsilon} \int_{-a/2}^{a/2} \left[ \frac{j}{\omega} \frac{dI(\acute{x})}{d\acute{x}} \right] H_0^{(2)}(k \mid x_{m-1/2} - \acute{x} \mid) d\acute{x}}_{\phi(x_{m-1/2})}$$

$$= -\vec{E}^{inc}(x)$$

(4.18)

We may now insert the finite difference expansion (4.16) into (4.18). The domain of the pulse functions reduces the limits of the integration and further splits the scalar potential terms. We can also insert (4.14) into the vector potential term of (4.18). This gives us:

$$-\frac{\omega\mu}{4} \int_{-a/2}^{a/2} \sum_{n=1}^{N} I_n P_n(\acute{x}) H_0^{(2)}(k \mid x - \acute{x} \mid) \, d\acute{x}$$

$$+\frac{1}{4j\epsilon} \int_{-a/2}^{a/2} \left[ \frac{j}{\omega} \sum_{n=1}^{N} I_n \left[ \frac{P_{n-1/2}(\vec{r}\,)}{\Delta x_{n-1/2}} - \frac{P_{n+1/2}(\vec{r}\,)}{\Delta x_{n+1/2}} \right] \right]$$

$$\cdot H_0^{(2)}(k \mid x_{m+1/2} - \acute{x} \mid) \, d\acute{x}$$

$$-\frac{1}{4j\epsilon} \int_{-a/2}^{a/2} \left[ \frac{j}{\omega} \sum_{n=1}^{N} I_n \left[ \frac{P_{n-1/2}(\vec{r}\,)}{\Delta x_{n-1/2}} - \frac{P_{n+1/2}(\vec{r}\,)}{\Delta x_{n+1/2}} \right] \right]$$

$$\cdot H_0^{(2)}(k \mid x_{m-1/2} - \acute{x} \mid) \, d\acute{x}$$

$$= -\vec{E}^{inc}(x_m) \tag{4.19}$$

We note:

$P_{n-1/2}(x)$ extends from $x_{n-1}$ to $x_n$

And:

$P_{n+1/2}(x)$ extends from $x_n$ to $x_{n+1}$

The pulse function's domain selects our integration limits. $I_n$ can be brought out front. We have already used pulse weighting to obtain the scalar potential terms. We must now weight the neglected vector potential term and the electric field term. Performing this operation results in (4.20)

$$\int_{x_{m-1/2}}^{x_{m+1/2}} -\hat{x} \bullet \vec{E}^{inc}(x) \, dx$$

$$= I_n \left\{ \int_{x_{m-1/2}}^{x_{m+1/2}} \hat{x} \bullet -\frac{\omega\mu}{4} \int_{x_{n-1/2}}^{x_{n+1/2}} H_0^{(2)}(k \mid x_m - \acute{x} \mid) \, d\acute{x} \, dx \right.$$

$$+ \frac{1}{4\omega\epsilon} \left[ \int_{x_{n-1}}^{x_n} \frac{H_0^{(2)}(k \mid x_{m+1/2} - \acute{x} \mid)}{\Delta x_{n-1/2}} \, d\acute{x} \right.$$

$$\left. - \int_{x_n}^{x_{n+1}} \frac{H_0^{(2)}(k \mid x_{m+1/2} - \acute{x} \mid)}{\Delta x_{n+1/2}} \, d\acute{x} \right]$$

$$-\frac{1}{4\omega\epsilon}\left[\int_{x_{n-1}}^{x_n}\frac{H_0^{(2)}(k\,|\,x_{m-1/2}-\acute{x}\,|)}{\Delta x_{n-1/2}}\,d\acute{x}\right.$$

$$\left.-\int_{x_n}^{x_{n+1}}\frac{H_0^{(2)}(k\,|\,x_{m-1/2}-\acute{x}\,|)}{\Delta x_{n+1/2}}\,d\acute{x}\right]\right\}\tag{4.20}$$

We can approximate the integration on the left–hand side of (4.20) by using the value of the electric field at the subdomain center (i.e., $x_m$) and multiplying by the subdomain length.

$$\int_{x_{m-1/2}}^{x_{m+1/2}}-\hat{x}\bullet\vec{E}^{inc}(x)\,dx$$

$$\approx-\hat{x}\bullet\vec{E}^{inc}(x_m)\,[x_{m+1/2}-x_{m-1/2}]$$

Or:

$$\approx-\hat{x}\bullet\vec{E}^{inc}(x_m)\,\Delta x_m$$

We also use this approximation for the vector potential term. When we do this, we obtain an expression for the matrix elements (4.21).

$$a_{mn}=-\frac{\Delta x_m\omega\mu}{4}\int_{x_{n-1/2}}^{x_{n+1/2}}H_0^{(2)}(k\,|\,x_m-\acute{x}\,|)\,d\acute{x}$$

$$+\frac{1}{4\omega\epsilon}\left[\int_{x_{n-1}}^{x_n}\frac{H_0^{(2)}(k\,|\,x_{m+1/2}-\acute{x}\,|)}{\Delta x_{n-1/2}}d\acute{x}\right.$$

$$\left.-\int_{x_n}^{x_{n+1}}\frac{H_0^{(2)}(k\,|\,x_{m+1/2}-\acute{x}\,|)}{\Delta x_{n+1/2}}d\acute{x}\right]$$

$$-\frac{1}{4\omega\epsilon}\left[\int_{x_{n-1}}^{x_n}\frac{H_0^{(2)}(k\,|\,x_{m-1/2}-\acute{x}\,|)}{\Delta x_{n-1/2}}d\acute{x}\right.$$

$$\left.-\int_{x_n}^{x_{n+1}}\frac{H_0^{(2)}(k\,|\,x_{m-1/2}-\acute{x}\,|)}{\Delta x_{n+1/2}}d\acute{x}\right]\tag{4.21}$$

The electric field terms are given by:

$$b_m=-E_0\Delta x_m\sin\phi_0 e^{-jk\cos\phi_0 x_m}\tag{4.22}$$

## 4.2  CALCULATION OF TWO–DIMENSIONAL RCS

Equation (3.32) is an expression for the RCS in three–dimensional space. Three–dimensional RCS is expressed as an area. We may also define RCS in two dimensions; in this case, RCS is expressed as a length. We can also define a logarithmic quantity with respect to a linear meter:

$$\sigma_{\text{dBlm}} = 10 \log_{10} \sigma \tag{4.23a}$$

RCS in two dimensions is defined mathematically as:

$$\sigma(\phi) = \lim_{r \to \infty} 2\pi r \frac{\mid E^{scat} \mid^2}{\mid E^{inc} \mid^2} \tag{4.23}$$

In two dimensions, the free–space Green's function is:

$$G(\vec{r}, \vec{r}) = \frac{1}{4j} H_0^{(2)}(k \mid \vec{r} - \vec{r} \mid)$$

The magnetic vector potential in two–dimensional space is:

$$\vec{A}(r) = \mu \int \int \vec{J}(\vec{r}) G(\vec{r}, \vec{r}) \, d\acute{s} \tag{4.24}$$

The electric field is given by:

$$\vec{E} = j\omega \vec{A} \tag{4.25}$$

Combining (4.23), (4.24), and (4.25) we obtain:

$$\vec{E}(r) = \frac{\omega\mu}{4} \int \int \vec{J}(\vec{r}) G(\vec{r}, \vec{r}) \, d\acute{s} \tag{4.26}$$

In the TM situation, the incident electric field along the strip is 1 V/m ($\mid \vec{E}_z^{inc} \mid^2 = 1$). The denominator of (4.23) is unity. This allows us to turn our attention to the numerator. To evaluate (4.26), we note that as $r \to \infty$, we can use the large argument approximation for the Hankle function:

$$H_0^{(2)}(r) \approx \sqrt{\frac{2}{\pi r}} e^{-j(r - \frac{\pi}{4})}$$

Substituting this into (4.26) and implementing (4.23) for the TM case, we obtain:

$$\sigma(\phi) = \frac{k\eta^2}{4} \left| \int_{strip} I(\acute{x}, \acute{y}) e^{jk(\acute{x} \cos\phi + \acute{y} \sin\phi)} \, d\acute{l} \right|^2 \tag{4.27}$$

In our case, the strip is restricted to the $x$-axis, which simplifies (4.27):

$$\sigma(\phi) = \frac{k\eta^2}{4} \left| \int_{-a/2}^{a/2} I(\hat{x}) e^{jk(\hat{x}\cos\phi)} \, d\hat{x} \right|^2 \tag{4.28}$$

For the TE situation, the formulation for RCS is performed in terms of the incident magnetic field $\vec{H}^{inc}$.[4] If we restrict the magnitude of the incoming electric field to 1 V/m, as we did in the TM case, the incoming magnetic field $\vec{H}$ is related by (4.29):

$$|H_z^{inc}| = \frac{|E^{inc}|}{\eta} \tag{4.29}$$

Under these conditions, equation (4.27) remains a valid expression of RCS in the TE strip case. In discrete form, (4.28) becomes:

$$\sigma(\phi) = \frac{k\eta^2}{4} \left| H \sum_{m=1}^{N} I_n e^{jk(x_m\cos\phi)} \right|^2 \tag{4.30}$$

## 4.3 NUMERICAL RESULTS OF TM AND TE SCATTERING

Solving for strip currents using (4.8) and (4.21), and (4.22) with a plane wave that is incident normal to the strip ($\phi_0 = 90°$), we obtain the current distribution in Figures 4.5 and 4.6. The current distribution is symmetrical, which we can expect from a symmetric incident wave.

The current of the TM solution becomes very large at the strip edges. We note that this is similar to the static situation, where charge becomes infinitely large at the edges. This large current density is a significant source of scattering from the strip.

At an edge the current in the TE solution has nowhere to go and thus must vanish. We see that the overall current magnitude is smaller for this polarization.

The RCS computations for a $\lambda/2$ metal strip in two dimensions extrapolate monotonically as seen in Table 4.1. $\sigma$ for the TM case is clearly 1.6797 and in the TE case 0.39839 appears to be a good estimate.

## 4.4 TM SCATTERING FROM RESISTIVE STRIPS

We next turn our attention to scattering from resistive strips. We will confine our discussion to TM scattering to maintain clarity, although this formulation may

Scattering From Conductive Strips

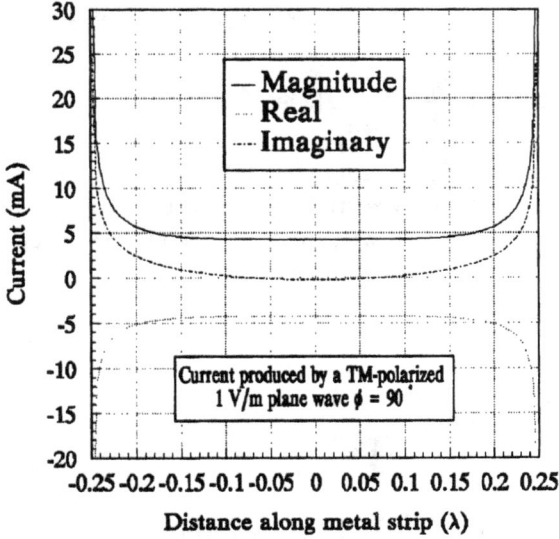

**Figure 4.5** The current distribution across a $\lambda/2$–wide perfectly conducting metallic strip created by a TM–polarized plane wave.

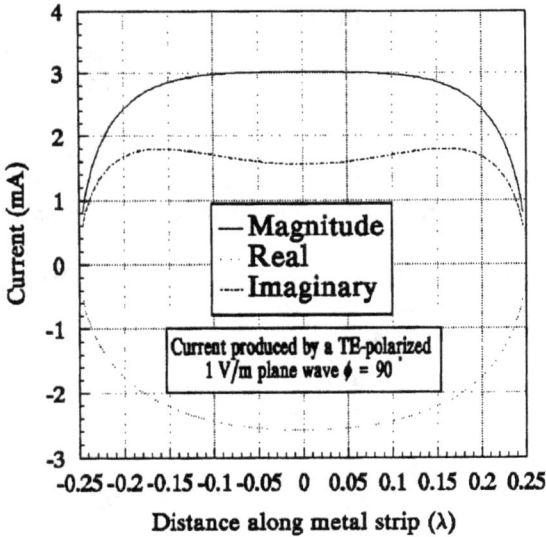

**Figure 4.6** The current distribution across a $\lambda/2$–wide perfectly conducting strip created by a TE–polarized plane wave.

Table 4.1

SCATTERING FROM PERFECTLY CONDUCTING INFINITE STRIP
LENGTH OF STRIP IN WAVELENGTHS:  0.5
INCIDENT ANGLE OF PLANE WAVE = 90 DEGREES

|  | TM SCATTERING | | TE SCATTERING | |
| M | (RCS) | EXTRAPOLATION | (RCS) | EXTRAPOLATION |
|---|---|---|---|---|
| 2 | 1.1533 | 1.1533 | .21268 | .21268 |
| 4 | 1.4098 | 1.6663 | .29353 | .37438 |
| 8 | 1.5440 | 1.6823 | .34541 | .40492 |
| 16 | 1.6120 | 1.6802 | .37253 | .39982 |
| 32 | 1.6459 | 1.6797 | .38577 | .39847 |
| 64 | 1.6629 | 1.6798 | .39218 | .39840 |
| 128 | 1.6713 | 1.6797 | .39531 | .39839 |
| 256 | 1.6755 | 1.6797 | .39686 | .39839 |
| 512 | 1.6776 | 1.6797 | .39758 | .39825 |

easily be extended to the TE case.* The surface resistance $R_s$ of a sheet of thin resistive material with a conductivity $\sigma_c$ and thickness $t$ is given by:

$$R_s = \frac{1}{\sigma_c t} \quad \Omega/\square \tag{4.31}$$

Sheets of resistive material with a very stable sheet resistance over large frequencies can be created using thin metallic films or thick film silkscreening techniques.[5] [6] [7] We note that the units of surface resistance are in ohms per square, the true unit is ohms. The square reminds us that it does not matter what size of a square, the measured sheet resistance is always the same.

The boundary condition at the surface of a thin resistive sheet is given in (4.32). This condition and (4.2) produces (4.33), which is an equation that governs TM scattering by a resistive strip:[8]

$$-E^{inc} = E^{scat} + R_s J(x) \tag{4.32}$$

$$-e^{jkx \cos \phi_0} = R_s I(x) + \frac{\omega\mu}{4} \int_{-a/2}^{a/2} I(\acute{x}) H_0^2(k \mid x - \acute{x} \mid)\, d\acute{x} \tag{4.33}$$

---

* Often TM scattering is the only polarization of interest in practical situations.

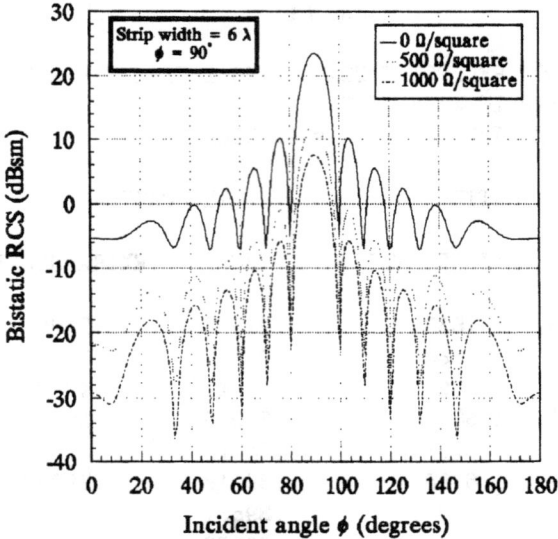

**Figure 4.7** The Bistatic RCS of a 6–$\lambda$ wide infinitely long resistive strip in two dimensions. The strip's RCS is calculated for sheet resistances of 0, 500, and 1000 $\Omega/\square$.

We use the same expansion scheme (pulses with point matching), as seen in Figure 4.2. The current is expanded as in (4.3).

$$I(\acute{x}) = \sum_{n=1}^{N} I_n P_n \tag{4.34}$$

In (4.33), we see that the difference in the incident and scattered field is related to the surface resistance and the current that is generated by the incoming field.

Substituting in the expansion scheme we obtain:

$$-e^{-jkx\cos\phi_0} = R_s I(x) + \frac{\omega\mu}{4} \sum_{n=1}^{N} I_n \int_{a_n}^{b_n} H_0^2(k\mid x - \acute{x}\mid)\, d\acute{x} \tag{4.35}$$

Rearranging and matching the field at $m$ points:

$$-e^{-jkx_m\cos\phi_0} = R_s I(x_m) + \frac{\omega\mu}{4} \sum_{n=1}^{N} I_n \int_{a_n}^{b_n} H_0^2(k\mid x_m - \acute{x}\mid)\, d\acute{x} \tag{4.36}$$

Where:

$$a_n = (n-1)H + H/2 - L/2$$

$$b_n = nH + H/2 - L/2$$

As before:

$$H = a/N$$

When $n = m$, then $I(x_m) \approx I_n$, which allows us to define matrix elements as:

$$b_m = -e^{-jkx_m \cos \phi_0} \tag{4.37}$$

$$a_{m,n} = \begin{cases} \frac{\omega\mu}{4} \int_{a_n}^{b_n} H_0^2(k \mid x_m - \acute{x} \mid) \, d\acute{x} & n \neq m \\ \frac{\omega\mu}{4} \int_{a_n}^{b_n} H_0^2(k \mid x_m - \acute{x} \mid) \, d\acute{x} + R_s & n = m \end{cases} \tag{4.38}$$

In Figure 4.7 RCS is plotted using the currents obtained with equations (4.37) and (4.38). The RCS for a 6–$\lambda$ wide strip of uniform sheet resistance $R_s$ of 0, 500, and 1000 $\Omega/\square$ is shown in Figure 4.7. We can see the pattern shapes are similar except for the magnitude of the reflected wave. As we can expect for an almost uniform distribution, the level of the first sidelobe is $\approx$ 13.0 dB down from the main lobe.

## 4.4.1   Quadratic Resistive Taper

The sheet resistance along a strip is not required to be a constant value. It is possible to vary the sheet resistance as a function of $x$. Changing the sheet resistance allows us to control aspects of the RCS pattern.

Antenna designers have a considerable amount of experience varying aperture distributions to obtain desired far–field radiation patterns. We can use this knowledge to shape the bistatic RCS pattern of a resistive strip.

The simplest taper we will examine is a quadratic taper expressed by (4.39)[9]

$$R_s(x) = 2\eta \left(\frac{x}{a}\right)^2 \ \Omega/\square \tag{4.39}$$

Where $\eta = 376.73 \ \Omega$.

Figure 4.8 shows the quadratic taper for a 6$\lambda$ strip. The quadratic taper reduces the first sidelobe to a level of -23 dB below the main lobe as illustrated in Figure 4.8. This taper has reduced the first sidelobes by 10 dB, compared with a uniform distribution. This is seen in Figure 4.9.

Scattering From Conductive Strips

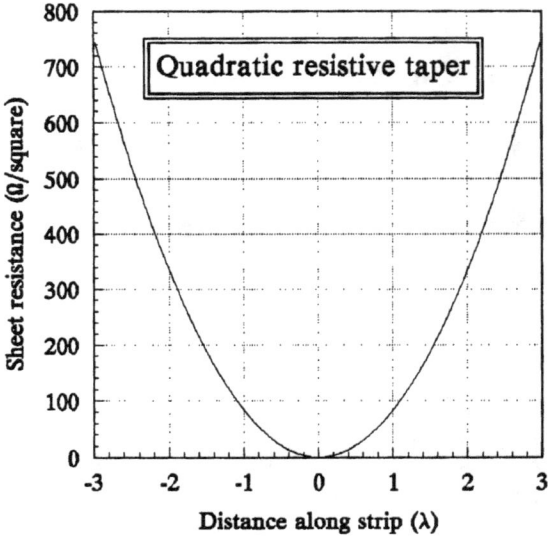

**Figure 4.8** A quadratic resistive taper creates an RCS pattern that has lower sidelobes than a sheet of constant resistance.

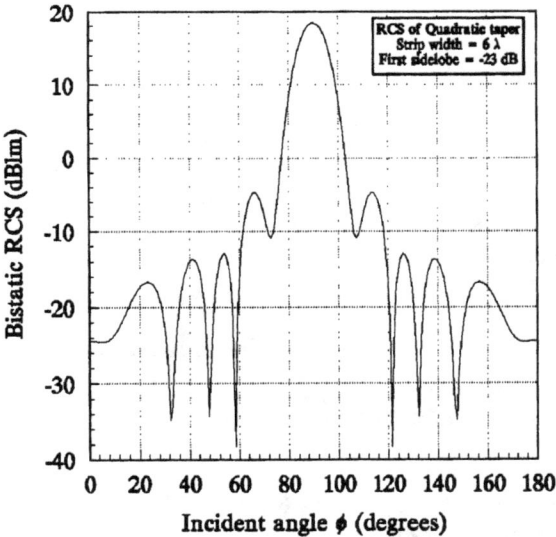

**Figure 4.9** This is the bistatic RCS pattern resulting from a plane wave at 90° incidence on the 6–λ strip of Figure 4.8.

## 4.4.2 Taylor Resistive Taper

We can see that it might be advantageous to control the RCS level of the main lobe and the sidelobes. The beamwidth of the main lobe is minimized when the RCS of the sidelobes all have the same level. When the sidelobes are all the same level, we have a Dolph-Chebyshev array distribution. This array distribution, which produces equal–level sidelobes, cannot be realized with a continuous distribution.

We note that a uniform distribution produces a set of sidelobes that decrease uniformly as seen in Figure 4.7. The bistatic RCS pattern for an incoming plane that is normal to the strip ($\phi_0 = 90°$) is proportional to $\sin \pi u/\pi u$. We can write this pattern function as an infinite product:

$$\frac{\sin \pi u}{\pi u} = \left[ \prod_{n=1}^{\infty} \left( 1 - \frac{u^2}{n^2} \right) \right] \tag{4.40}$$

Each time $\sin \pi u/\pi u$ passes through a zero we have a pattern null. We can manipulate the pattern zeros to synthesize a desired radiation pattern. A finite number of zeros may be removed from (4.40) and replaced with a set of zero locations that correspond to a *Dolph-Chebyshev array distribution.*

The number of array elements $\bar{n}$ determines the number of zero pairs to be manipulated: $(\bar{n} - 1)$. This situation produces sidelobes of a prescribed level below the main beam which decrease in a $\sin x/x$ manner thereafter. This distribution is known as a *Taylor distribution.* We can use this distribution to synthesize a resistive taper that has sidelobes a specified number of decibels below the main beam.[10]

In equation (4.41), the first $\bar{n} - 1$ pairs of zeros of the sinc function are removed by the denominator. The zeros of the Dolph-Chebyshev array distribution that replace them are the $u_n$'s found in the numerator:

$$S(u) = \frac{\sin \pi u}{\pi u} \frac{\prod_{n=1}^{\bar{n}-1} (1 - u^2/u_n^2)}{\prod_{n=1}^{\bar{n}-1} (1 - u^2/n^2)} \tag{4.41}$$

Using (4.40) we obtain:

$$S(u) = \left[ \prod_{n=1}^{\bar{n}-1} \left( 1 - \frac{u^2}{u_n^2} \right) \right] \left[ \prod_{n=\bar{n}}^{\infty} \left( 1 - \frac{u^2}{n^2} \right) \right] \tag{4.42}$$

The values of $u_n$ for a Dolph–Chebyshev array are given by (4.43):

$$u_n = \bar{n} \left[ \frac{A^2 + (n - \frac{1}{2})^2}{A^2 + (\bar{n} - \frac{1}{2})^2} \right]^{\frac{1}{2}} \tag{4.43}$$

$A$ is a measure of the sidelobe level on either side of the main beam. $A$ is defined by:

$$A = \frac{\cosh^{-1} b}{\pi} \qquad (4.44)$$

where $b$ is the ratio of the main beam to the first sidelobe and is related to the sidelobe level (SLL) in dB by:

$$\text{SLL} = 20 \log (b) \qquad (4.45)$$

We have the desired pattern function in (4.42). The aperture distribution that produces it may be found by expanding it with a cosine series and using the Fourier transform. The two expressions are equated and matched at integer values of $u$. When this is done we obtain for the case of our strip.[11]

$$g(x) = \frac{e^{-j\beta x}}{2a} \left[ S(0) + 2 \sum_{m=1}^{\bar{n}-1} S(m) \cos \frac{m\pi x}{a} \right] \qquad (4.46)$$

The sheet resistance as a function of $x$ is related to distribution $g(x)$ by:

$$R_s(x) = \frac{\eta}{g(x)} - \frac{\eta}{2} \ \Omega/\square \qquad (4.47)$$

As an example, we will use (4.47) to synthesize a resistive strip with an aperture distribution for a Taylor 30/30 sidelobes (30 dB sidelobes on either side of the main lobe) with $\bar{n} = 8$.

Figure 4.10 presents the sheet resistance required to produce a -30 dB sidelobe level for the scattered field from the strip at normal incidence. Note the similarity to the quadratic taper with the exception of either edge of the strip. The taper becomes approximately constant at the edges. This is a pedestal on top of which the Dolph–Chebyshev distribution resides. It maintains the $1/u$ reduction in sidelobe level as one is far away from the main beam. In Figure 4.11 the calculated bistatic sidelobe level is seen to be -30 dB as expected.

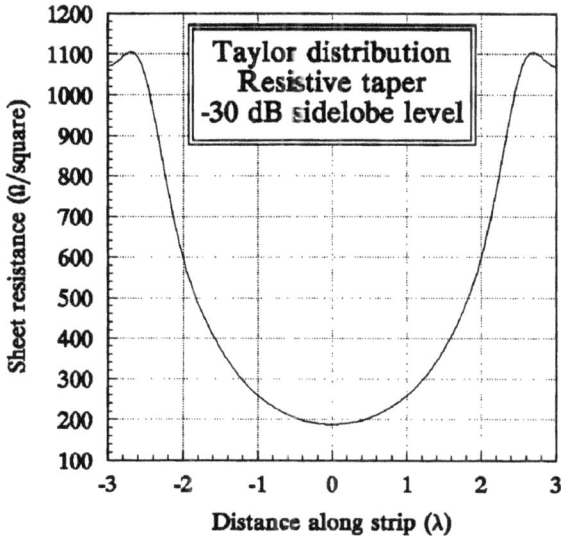

**Figure 4.10**  A Taylor resistive taper with $\bar{n} = 8$ to synthesize a -30 dB sidelobe level.

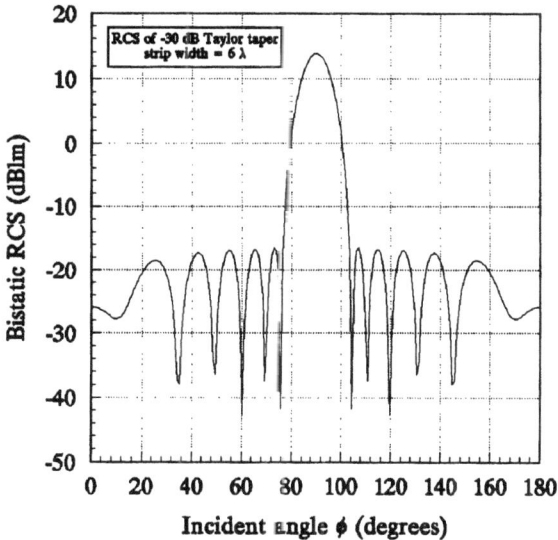

**Figure 4.11**  This is the bistatic RCS pattern resulting from a plane wave at 90° incidence on the 6 $\lambda$ strip of Figure 4.10. We note the sidelobe level is -30 dB.

# References

[1] Glisson, A.W. and Wilton, D.R., "Simple and Efficient Numerical Methods for Problems of Electromagnetic Radiation and Scattering From Surfaces," *IEEE Transactions on Antennas and Propagation*, Vol. AP – 28, No. 5, September 1980, pg. 593–603.

[2] Harrington, Roger F., *Time Harmonic Electromagnetic Fields*, McGraw Hill, 1961, pg. 198–204.

[3] Sadiku, Matthew N.O., *Numerical Techniques in Electromagnetics*, CRC Press, 1992, pg. 317–320.

[4] Harrington, Roger F., *Field Computation by Moment Methods*, Malabar, Florida, Robert E. Krieger Publishing Company, 1968 (1983 reprint edition), pg. 53–54.

[5] Hanson, R.C. and Pawlewicz, W.T., "Effective Conductivity and Microwave Reflectivity of Thin Metallic Films," *IEEE Transactions on Microwave Theory and Techniques*, Vol. 30, No. 11, November 1982, pg. 2064–2066.

[6] Bancroft, Randy, *Resistive Sheet Development*, Ball Aerospace System Engineering Report, July 1990, Broomfield, Colorado.

[7] Bancroft, Randy, *Design of Tapers by Use of Open and Short-Circuit Resistive Grids*, Ball Aerospace SER.3484.0020.U.DC, October 1989, Broomfield, Colorado

[8] Senior, Thomas B.A., "Backscattering from Resistive Strips," *IEEE Transactions on Antennas and Propagation*, Vol. AP-27, No. 6, November 1979.

[9] Senior, Thomas B.A. & Liepa, V.V., "Backscattering From Tapered Resistive Strips," *IEEE Transactions on Antennas and Propagation*, Vol. AP-32, No. 7, July 1984, pg. 747–751.

[10] Haupt, Randy L. and Liepa, Valdis V., "Synthesis of Tapered Resistive Strips," *IEEE Transactions on Antennas and Propagation*, Vol. AP-35, No. 11, November 1987.

[11] Milligan Thomas, *Modern Antenna Design* McGraw–Hill 1985 pg. 141–144.

# Chapter 5

# Scattering From Two–Dimensional Contours

## 5.1  RCS OF PERFECTLY CONDUCTING TWO–DIMENSIONAL CONTOUR

### 5.1.1  TM Polarization

In this chapter we will concern ourselves with two–dimensional scattering from an arbitrary closed contour. A $\hat{z}$ polarized incident plane wave (Figure 5.1) will induce surface current $I_z$ along the $z$ axis. The currents in turn produce a scattered electric field $E_z^s$. The condition at the contour boundary is:

$$E_z = E_z^{inc} + E_z^{scat} = 0 \text{ on C} \tag{5.1}$$

The integral equation that describes the electric field resulting from a current distribution along the contour is given by[1]

$$E_z(\vec{r}) = -\frac{\omega\mu}{4} \oint_C I_z(\vec{r})H_0^{(2)}(k \mid \vec{r} - \vec{r} \mid) \, d\hat{l} \tag{5.2}$$

Using equation (5.1) with (5.2) and restricting $\vec{r}$ to be on contour C we obtain:

$$E_z^{inc}(\vec{r}) = \frac{\omega\mu}{4} \oint_C I_z(\vec{r})H_0^{(2)}(k \mid \vec{r} - \vec{r} \mid) \, d\hat{l} \qquad [\vec{r} \text{ on C}] \tag{5.3}$$

Before using (5.3) to implement the moment method, we will write it as a function of polar angle $\theta$. The current is represented using pulse–expansion functions along the contour. These are defined below:

$$P_n(\theta) = \begin{cases} 1 & \theta \in [\theta_n, \theta_{n+1}] \\ 0 & \text{elsewhere} \end{cases} \tag{5.4}$$

$$I_z(\vec{r}(\theta)) = \sum_{n=1}^{N} P_n(\theta)I_n$$

Rewriting (5.3) as a function of polar angle and inserting (5.4) we obtain (5.5).

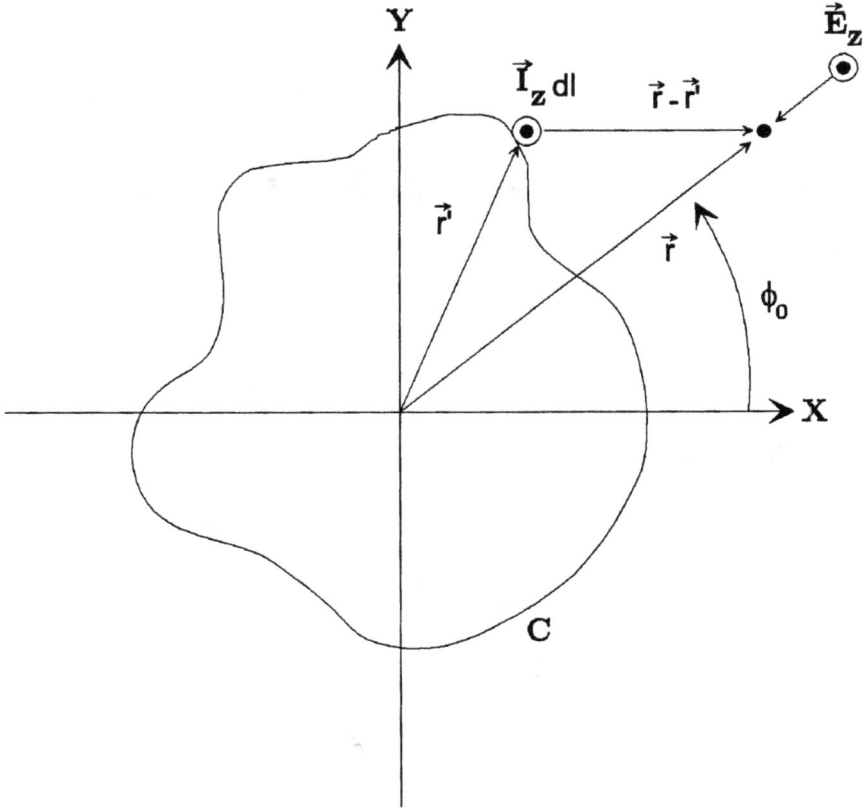

**Figure 5.1** A cylindrical perfectly conducting contour of constant cross–section and extending infinitely in the $\pm z$ direction, has an incoming plane wave produce a current along the two–dimensional contour. The contour is described by the vector $\vec{r}'$. The observation point is given by $\vec{r}$.

$$E_z^{inc}(\vec{r}(\theta)) = \frac{\omega\mu}{4} \oint_C \sum_{n=1}^{N} P_n(\acute{\theta}) I_n H_0^{(2)}(k \mid \vec{r}(\theta) - \vec{r}(\acute{\theta}) \mid) \mid \vec{r}(\acute{\theta}) \mid d\acute{\theta} \qquad (5.5)$$

The pulse functions select out the integration limits:

$$E_z^{inc}(\vec{r}(\theta)) = I_n \sum_{n=1}^{N} \frac{\omega\mu}{4} \int_{\theta_n}^{\theta_{n+1}} H_0^{(2)}(k \mid \vec{r} - \vec{r}(\acute{\theta}) \mid) \mid \vec{r}(\acute{\theta}) \mid d\acute{\theta}$$

Where: $d\acute{l} = |\acute{r}| \ d\acute{\theta}$

We can now point–match the electric field at $M$ points to obtain as many equations as unknowns:

$$E_z^{inc}(\vec{r}(\theta_m)) = I_n \sum_{n=1}^{N} \frac{\omega\mu}{4} \int_{\theta_n}^{\theta_{n+1}} H_0^{(2)}(k \mid \vec{r}(\theta_m) - \vec{r}(\acute{\theta}) \mid) \mid \vec{r}(\acute{\theta}) \mid \ d\acute{\theta} \qquad (5.6)$$

For computational purposes it is convenient to write (5.6) as:

$$E_z^{inc}(\vec{r}(\theta_m)) = I_n \frac{\omega\mu}{4} \sum_{n=1}^{N} \int_{(n-1)\Delta\theta}^{n\Delta\theta} H_0^{(2)}(k \mid \vec{r}(\theta_m) - \vec{r}(\acute{\theta}) \mid) \mid \vec{r}(\acute{\theta}) \mid \ d\acute{\theta} \qquad (5.7)$$

Where:
$\Delta\theta = \frac{2\pi}{N}$ and $n = 1, 2, 3, \dots, N$

And:
$\theta_m = (m - 1)\Delta\theta + \frac{\Delta\theta}{2}$ $\quad m = 1, 2, 3, \dots M$

Equation (5.7) allows us to identify the impedance matrix as:

$$a_{mn} = \frac{\omega\mu}{4} \int_{(n-1)\Delta\theta}^{n\Delta\theta} H_0^{(2)}(k \mid \vec{r}(\theta_m) - \vec{r}(\acute{\theta}) \mid) \mid \vec{r}(\acute{\theta}) \mid \ d\acute{\theta} \qquad (5.8)$$

The incident electric field provides our excitation matrix $b$:

$$b_m = E_z^{inc}(\vec{r}(\theta_m)) = e^{jk(x_m \cos \phi_0 + y_m \sin \phi_0)} \qquad (5.9)$$

Where:

$x_m = \mid r(\theta_m) \mid \cos \theta_m$
$y_m = \mid r(\theta_m) \mid \sin \theta_m$

Solving for the unknown currents along the contour allows us to evaluate the two–dimensional bistatic RCS from a contour of any shape defined by $\vec{r}(\theta)$. Equation (4.27) remains a valid equation for the calculation of RCS in the case of a closed contour for the TM case. It is reproduced as:

$$\sigma(\phi) = \frac{k\eta^2}{4} \left| \oint I(\acute{x}, \acute{y}) e^{jk(\acute{x} \cos \phi + \acute{y} \sin \phi)} \ d\acute{l} \right|^2 \qquad (5.10)$$

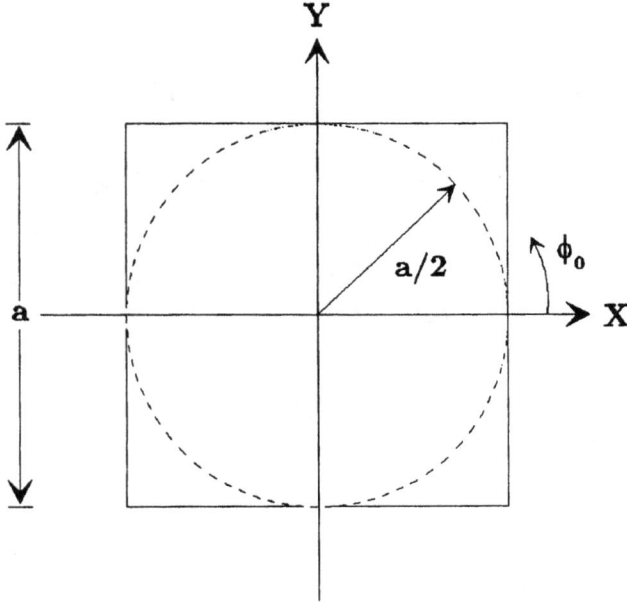

**Figure 5.2** Two different contours are examined in the TM scattering case: a circular contour and a square contour. The current and bistatic RCS from each will be evaluated.

In discreet form it becomes:

$$\sigma(\phi) = \frac{k\eta^2}{4} \left| H \sum_{m=1}^{N} I_n e^{jk(x_m \cos\phi + y_m \sin\phi)} \right|^2 \qquad (5.11)$$

## 5.1.2 Numerical Results for TM Scattering

To examine the effect contour shape has on RCS we choose a square contour and a circular contour for solution (Figure 5.2). The circular contour has a diameter equal to the length of each side of the square contour. We choose $\lambda/2$ for the value of $a$. Recall that we are calculating the two–dimensional RCS from two infinitely long pipes–one has a square contour, and the other, a circular contour.

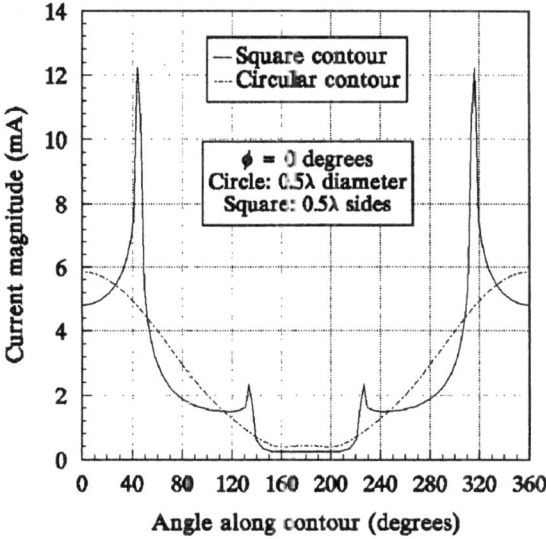

**Figure 5.3** This is a plot of the current generated on a square contour and a circular contour. The TM plane wave is incident at $0°$.

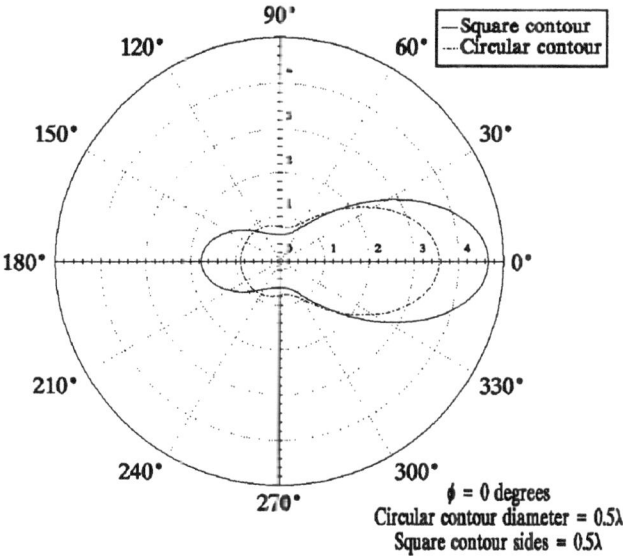

**Figure 5.4** This is a plot of bistatic RCS from currents of the circular and square contours with $\phi_0 = 0°$. Note that the RCS of the square is much larger in both backscatter and forward scatter.

In Figures 5.3 and 5.4 we consider a 1V/m plane wave incident at $\phi_0 = 0°$. Note that at 0° the current is about 4.8 mA for the square contour. The circular contour current is nearly 6 mA. As we move along the contour, we see that the current peaks sharply at each corner of the square contour. The circular contour does not have these sharp corners; it has instead a gentle sinusoidal current distribution. The RCS $\sigma/\lambda$ is about 4.5 at 0° ($\phi = 0°$) for the square. The circle is lower at $\sigma/\lambda \approx 3.5$.

The angle of the incoming plane wave is next moved to $\phi_0 = 45°$. We present the results for this case in Figures 5.5 and 5.6. Note that the largest current spike is at 45°. This is where the first corner is encountered by the plane wave. The two corners on either side have equal current and the shadowed corner in back has the smallest induced current.

The bistatic polar plot of RCS (Figure 5.6) shows a bifurcated backlobe. Each lobe is created by the pair of equivalent current spikes at 135° and 315°. Once again, the backscatter from the square contour is larger than that of the circular.

### 5.1.3  TE Polarization

In the case of TE polarization, (Figure 5.7), the $\vec{H}$ field vector is parallel with the z–axis. At the contour surface, the $H$ field satisfies the Neumann boundary condition ($\frac{\partial \phi}{\partial n}(\vec{r}) = 0$). With the appropriate boundary conditions and allowing the observation point to approach the contour surface, we obtain[2]

$$H_z^{inc}(\vec{r}) = \frac{1}{2}I_t(\vec{r}) + \lim_{\Delta \to 0} \oint_{C-\Delta} \frac{\partial G(\vec{r},\acute{r})}{\partial \acute{n}} I_t(\acute{r}) \ d\acute{C} \qquad (5.12)$$

Where $\Delta$ is a small segment of $C$ containing the point $\vec{r}$. $\hat{n}$ is an outward–pointing unit vector perpendicular to the contour at each source point. $G(\vec{r},\acute{r})$ is the two–dimensional free–space Green's function. $I_t$ is the tangential current flow around the contour.

We use the same pulse functions as in the TM case.

$$P_n(\theta) = \begin{cases} 1 & \theta \in [\theta_n, \theta_{n+1}] \\ 0 & \text{elsewhere} \end{cases} \qquad (5.13)$$

$$I_n(r(\theta)) = \sum_{n=1}^{N} P_n(\theta)I_n \qquad (5.14)$$

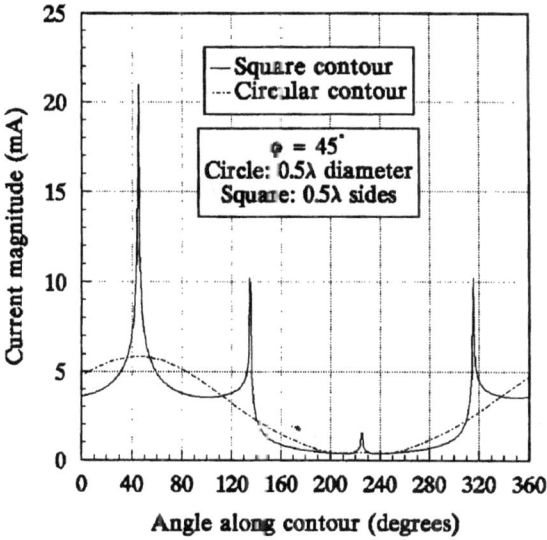

**Figure 5.5** This is a plot of the current generated on a square contour and a circular contour. The TM wave is incident at 45°.

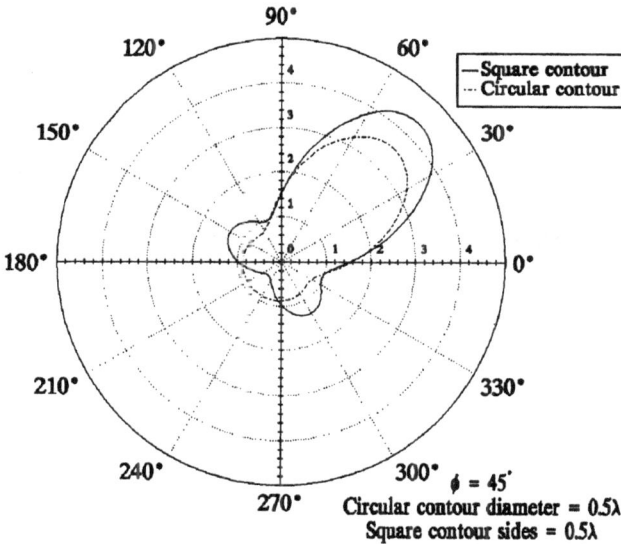

**Figure 5.6** This is a plot of bistatic RCS from currents of the circular and square contours with $\phi_0 = 45°$. We note that the RCS of the square is much larger in both backscatter and forward scatter.

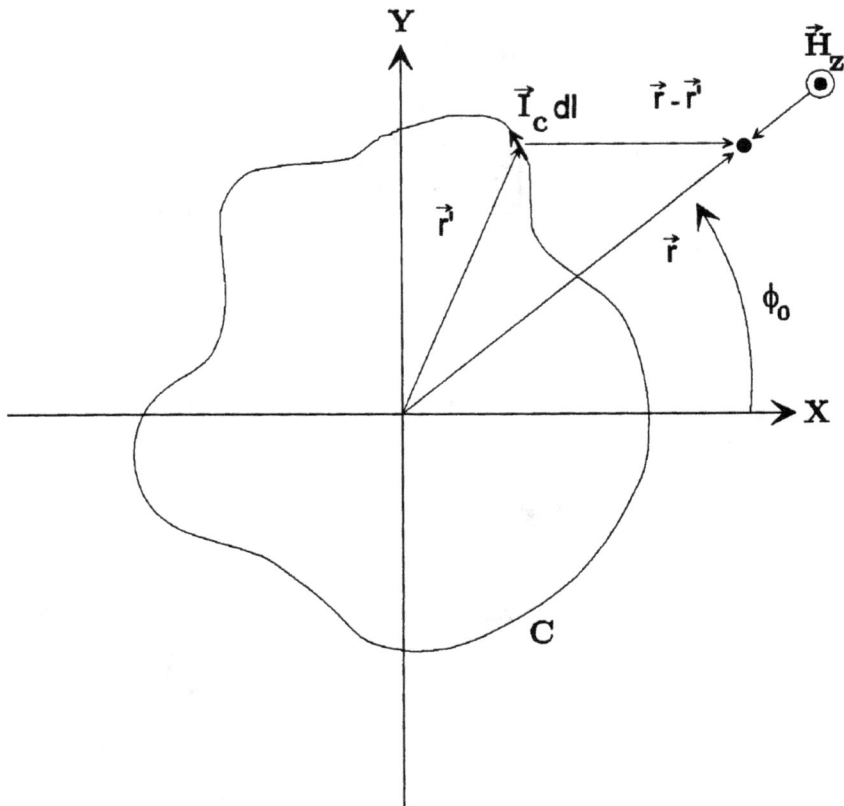

**Figure 5.7** A cylindrical contour of constant cross–section, extending infinitely in the $\pm z$ direction, has an incoming plane wave produce a current around the two–dimensional contour. The contour is described by the vector $\vec{r}$. The observation point is given by $\vec{r}$.

In terms of $\theta$

$$H_z^{inc}(\vec{r}(\theta)) = \frac{1}{2} I_t(\vec{r}(\theta))$$

$$+ \lim_{\Delta \to 0} \oint_{C-\Delta} I_n \sum_{n=1}^{N} P_n(\theta) \frac{\partial G(\vec{r}(\theta), \vec{r}(\acute{\theta}))}{\partial \acute{n}} \mid \vec{r}(\acute{\theta}) \mid d\acute{\theta} \qquad (5.15)$$

The pulse functions determine the integration limits. We can then match equation (5.15) at $M$ match angles $\theta_m$.

$$H_z^{inc}(\vec{r}(\theta_m)) = \frac{1}{2}I_m(\vec{r}(\theta_m))$$

$$+I_n \sum_{n=1}^{N} \int_{\theta_n}^{\theta_{n+1}} \frac{\partial G(\vec{r}(\theta_m), \vec{r}(\acute{\theta}))}{\partial \acute{n}} \mid \vec{r}(\acute{\theta}) \mid d\acute{\theta} \qquad (5.16)$$

The match point current $I_m$ is equal to $I_n$ when $(n = m)$. This allows us to write our impedance matrix [A] as:

When $m \neq n$:

$$a_{mn} = \int_{\theta_n}^{\theta_{n+1}} \frac{\partial G(\vec{r}(\theta_m), \vec{r}(\acute{\theta}))}{\partial \acute{n}} \mid \vec{r}(\acute{\theta}) \mid d\acute{\theta} \qquad (5.17a)$$

When $m = n$

$$a_{nn} = \frac{1}{2} + \int_{\theta_n}^{\theta_{n+1}} \frac{\partial G(\vec{r}(\theta_n), \vec{r}(\acute{\theta}))}{\partial \acute{n}} \mid \vec{r}(\acute{\theta}) \mid d\acute{\theta} \qquad (5.17b)$$

and:

$$b_m = H_z^{inc}(\vec{r}(\theta_m)) = e^{jk(x_m \cos \phi_0 + y_m \sin \phi_0)} \qquad (5.18)$$

or

$$b_m = H_z^{inc}(\vec{r}(\theta_m)) = e^{jk(|r(\theta_m)|[\cos \theta_m \cos \phi_0 + \sin \theta_m \sin \phi_0])} \qquad (5.19)$$

We can now solve for the unknown currents. The difference in this case is the derivative acting upon the Green's function in the integrand. One way to evaluate the integrand is to use a numerical derivative as demonstrated in Chapter 1. A numerical derivative with Richardson's extrapolation is very accurate in this case and produces good results. Closed–form expressions are also presented in the literature.[3]

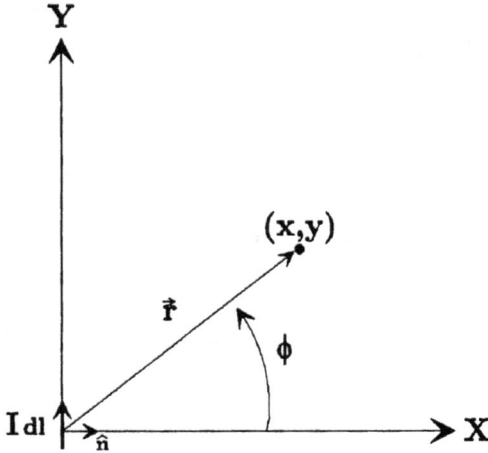

**Figure 5.8** A differential current element in the $x$–$y$ plane

## 5.1.4 Calculation of RCS in the TE Case

To calculate the bistatic RCS due to the TE–generated contour current, we first consider a differential size current element in the $x$–$y$ plane. This is illustrated in Figure 5.8.

A differential current behaves as a point source. The vector potential from the differential current element is:

$$dA_y = \frac{\mu}{4j} H_0^{(2)}(kr) I dl \tag{5.20}$$

The magnetic field is related to the vector potential by:

$$\vec{H} = \frac{1}{\mu} \nabla \times \vec{A} \tag{5.21}$$

Which gives:

$$dH_z = \frac{\partial}{\partial x} \left( \frac{\mu}{4j} H_0^{(2)}(kr) I dl \right) \tag{5.22}$$

$$dH_z = \frac{jk}{4} \cos\phi H_1^{(2)}(kr) I dl \tag{5.23}$$

For an arbitrary origin:

$$dH_z = \frac{jk}{4}(\hat{n} \bullet \hat{R})H_1^{(2)}(k \mid \vec{r} - \vec{r}_n \mid)Idl \qquad (5.24)$$

Where $\hat{n}$ is a unit vector perpendicular to the contour. $\hat{R}$ is a unit vector pointing from the source point at $\vec{r}_n$ to $\vec{r}$, which is an arbitrary field point:

$$\hat{R} = \frac{\vec{r} - \vec{r}_n}{\mid \vec{r} - \vec{r}_n \mid} \qquad (5.25)$$

Integrating, we obtain the field:

$$H_z(\vec{r}) = \frac{jk}{4} \oint_C (\hat{n} \bullet \hat{R})H_1^{(2)}(k \mid \vec{r} - \vec{r}_n \mid)I(\vec{r}_n) \; d\acute{C} \qquad (5.26)$$

We define the RCS for the TE case as:

$$\sigma(\phi) = \lim_{r \to \infty} 2\pi r \frac{\mid \vec{H}^s(\phi) \mid^2}{\mid \vec{H}^{inc} \mid^2} \qquad (5.27)$$

$\vec{H}^s(\phi)$ is the scattered field at a large distance from the source current. We obtain this by using the large argument expression for the Hankel function:

$$H_1^{(2)}(kr) = \sqrt{\frac{2}{\pi kr}}e^{-j(kr-3\pi/4)} \qquad (5.28)$$

Using this expression in (5.26), we obtain:

$$H_z^s(\phi) = \frac{jk}{4}\sqrt{\frac{2}{\pi kr}} \int_C I(\acute{x}, \acute{y})(\hat{n} \bullet \hat{R})e^{jk(\acute{x}\cos\phi_0 + \acute{y}\sin\phi_0)} \; d\acute{C} \qquad (5.29)$$

Which, using the RCS definition (5.27), gives:

$$\sigma(\phi) = \frac{k}{4}\left| \int_C I(\acute{x}, \acute{y})(\hat{n} \bullet \hat{R})e^{jk(\acute{x}\cos\phi_0 + \acute{y}\sin\phi_0)} \; d\acute{C}\right|^2 \qquad (5.30)$$

Note the similarity to the calculation of RCS for the TM situation. The major difference is the $\hat{n} \bullet \hat{R}$ term of the integrand. When the differential currents are along the contour, we must take their element factors into account. In the TM situation, we see the pattern factor is in the $x$-$z$ and $y$-$z$ planes and therefore does not appear.

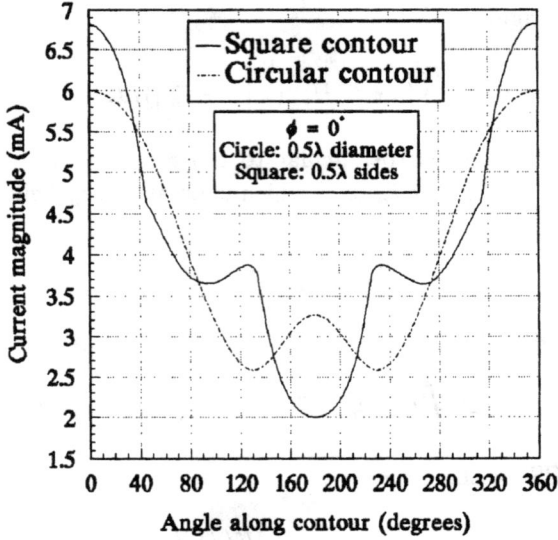

**Figure 5.9** This is a plot of the current generated on a square contour and a circular contour. The TE plane wave is incident at 0°.

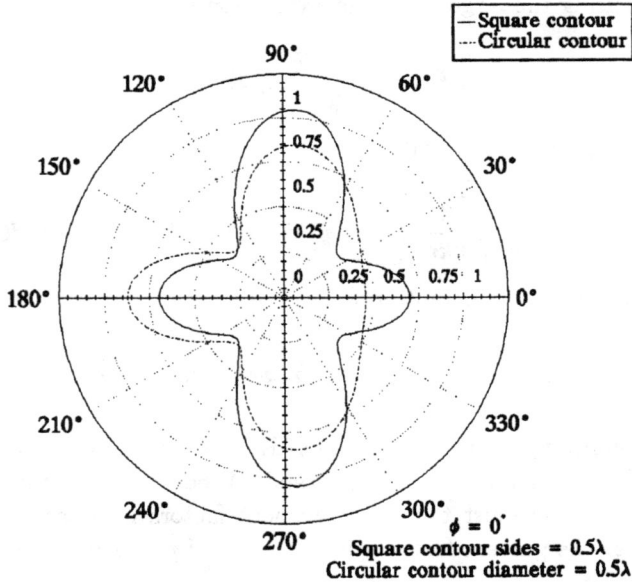

**Figure 5.10** This is a plot of bistatic RCS from currents of the circular and square contours with $\phi_0 = 0°$. We note that the RCS of the square is much larger in both back–scatter and forward scatter.

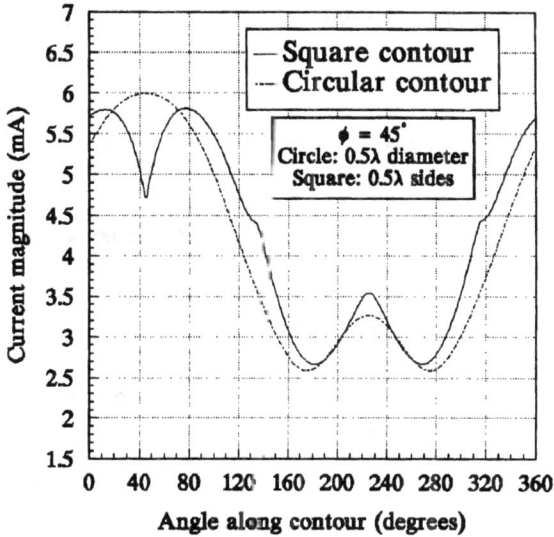

**Figure 5.11** This is a plot of the current generated on a square contour and a circular contour. The TE plane wave is incident at 45°.

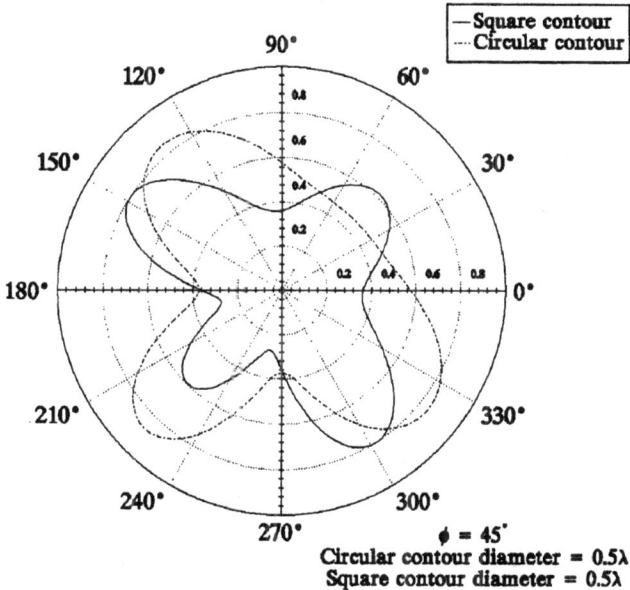

**Figure 5.12** This is a plot of bistatic RCS from currents of the circular and square contours with $\phi_0 = 45°$. We note that the RCS of the square is much larger in both back–scatter and forward scatter.

### 5.1.5  Numerical Results of TE Scattering From a Contour

Figure 5.9 shows scattering results for circular and square contours in the TE case. The circular contour has a diameter of $0.5\lambda$, and each side of the square contour is also $0.5\lambda$. We can see from the plot (Figure 5.9), that the current on the circular contour smoothly decreases as we progress from the incident angle. It slightly increases on the side opposite the incident angle. The square contour, by contrast, begins with a larger current at the angle of incidence. At 45°, we see a kink in the current corresponding to a corner and a corresponding change in slope. The next corner at 135° shows another substantial change.

The RCS plot (Figure 5.10) reveals that the scattering back toward incidence is lower than that scattered perpendicular and in the forward direction. This situation is more pronounced for the circular contour than for the square.

The current produced on the contour with a plane wave incident angle of 45° is plotted in Figure 5.11. It is interesting to note the current minimum at the corner of the square, which is coincident with the incident wave at 45°. We see a current change similar to that in Figure 5.9 at 135°. A local current maximum is seen for the square contour on the corner opposite of the incident angle. By coincidence, the local maximum of the circular contour coincides with the square maximum.

The RCS plot of the current (Figure 5.12) shows that the backscatter at the 45° incident angle is similar to that at 0°. The perpendicular scattering is tilted forward.

### 5.2  Monostatic and Bistatic RCS

So far we have concentrated on calculating the bistatic RCS of an object. To do this we have a plane wave at an incident angle $\phi_0$ impress currents on a two–dimensional contour. It is then possible to calculate the radiated field in any $\phi$ direction from the currents. The fields are then used to calculate the bistatic RCS.

This situation is mirrored by taking a single transmitting antenna and using it to illuminate an object. The object then has currents flowing on its exterior. A second (receiving) antenna may be moved through an angle of 0° to 360° at a constant radius with its main beam pointing toward the object, which is at the center of our test–area radius. The second antenna measures the electric field from the object at each angle. The characteristics of the transmitting antenna allow us to obtain the incident field strength. We can then calculate the bistatic RCS from the measured values.

Often, we are confined to using a single antenna for both transmission and reception. This is the *monostatic* RCS of an object. In this case, we take an antenna and illuminate an object with it. At the same angle, we measure the reflected wave and calculate RCS.

To calculate the monostatic RCS using the moment method, we would choose the angle of the incoming plane wave, solve for the currents, and then calculate the electric field and RCS at that single angle. We then move to the next angle of interest and repeat the procedure.

With the moment method, we solve a matrix equation of the type

$$[a][I] = [b] \qquad (5.31)$$

for the current values $[I]$.

Once we have calculated the value of the $[a]$ matrix, we can generate a $[b]$ matrix that corresponds to a given incident angle. We may then solve (5.31) for the currents at that incident angle and the RCS at that angle calculated. The $[a]$ matrix values do not change, and only a new $[b]$ matrix is required to solve (5.31) for the next set of currents produced by a new incident angle.

The same $0.5\lambda$ square and circular contours previously used for illustrating TM and TE scattering from square and circular contours have their monostatic RCS plotted in Figures 5.13 and 5.14.

The TM case shows that, as one would expect for a circular contour, the reflected wave at any incident angle should be constant. The square contour has a larger overall RCS with the RCS of the sides larger than that of the corners.

The TE case shows again that the circular contour has a monostatic RCS independent of incident angle. The square contour has a very pronounced oscillation with incident angle.

It is instructive to see the monostatic RCS of the two contours with the bistatic RCS at a number of angles. In the TM case, Figure 5.15, we clearly see each of the bistatic plots contained within the monostatic contour. In this case, the bistatic RCS is smaller overall than the monostatic RCS at any angle. The TE case is quite different: the monostatic RCS is much smaller than the maximum values of the bistatic RCS that makes it up.

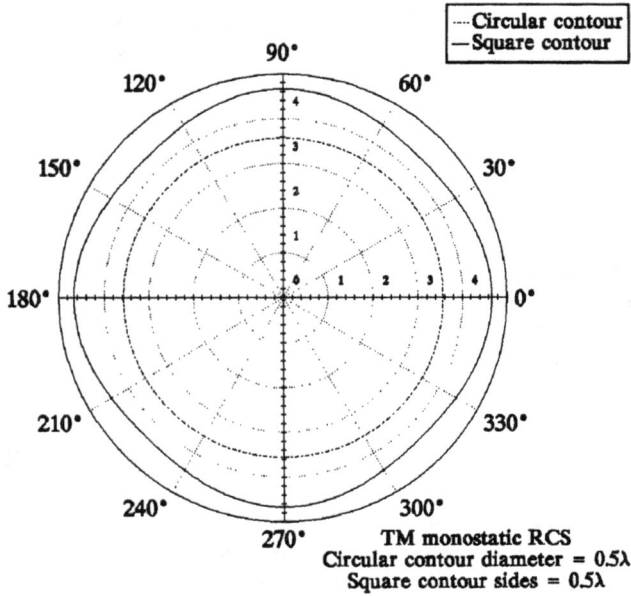

**Figure 5.13** The TM monostatic radar cross section of square and circular contours. The circular contour presents a constant monostatic RCS. The square reveals its sides and corners.

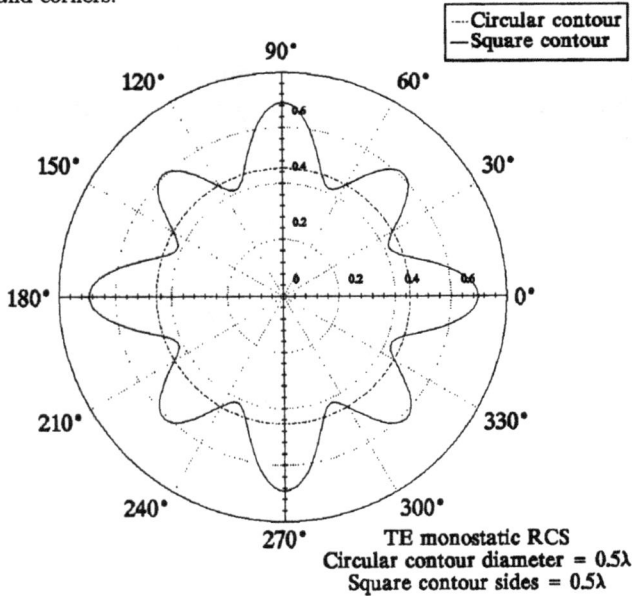

**Figure 5.14** The TE monostatic RCS of square and circular contours. The circular contour presents a constant monostatic RCS. The square reveals its sides and corners.

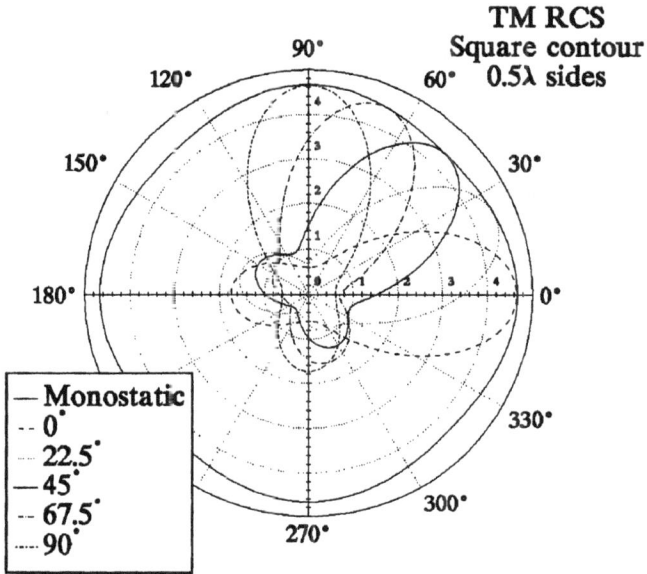

**Figure 5.15** The TM monostatic RCS and the bistatic RCS that forms it.

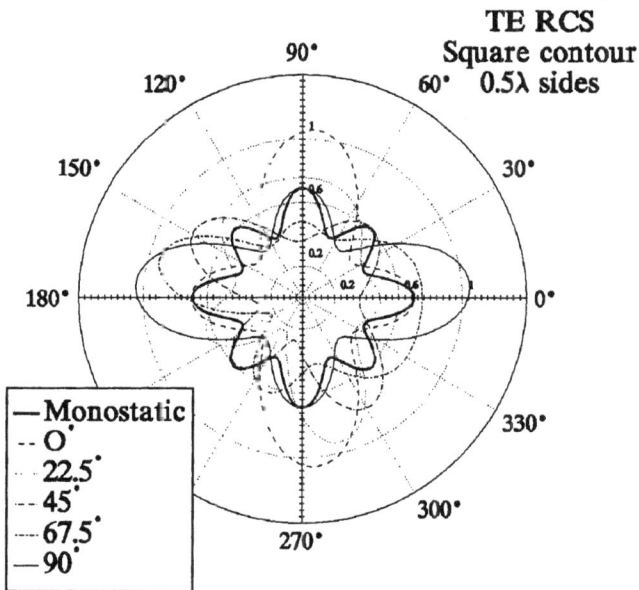

**Figure 5.16** The TE monostatic RCS and the bistatic RCS that forms it.

## References

[1] Harrington, Roger F., *Field Computation by Moment Methods*, Malabar, Florida, Robert E. Krieger Publishing Company, 1968 (1983 reprint edition), pg. 42.

[2] Mei K.K., and Van Bladel, J. G., "Scattering by Perfectly-Conducting Rectangular Cylinders," *IEEE Transactions on Antennas and Propagation*, March 1963, pg. 185–192. See also Comments on "Scattering by Conducting Rectangular Cylinders," *IEEE Transactions on Antennas and Propagation*, March 1964, pg. 235–236.

[3] Mittra, R. ed., *Computer Techniques for Electromagnetics*, Oxford, Pergamon Press, 1973. pg. 173–174.

# Chapter 6

# Radar Cross Section of a Flat Plate

## 6.1 RCS OF A THIN, PERFECTLY CONDUCTING SQUARE PLATE

### 6.1.1 Moment Method Solution (Pulse/Pulse)

In this chapter, we concern ourselves with scattering from a thin, perfectly conducting metal plate (Figure 6.1). The electric field and current on the plate are related by a pair of coupled scalar equations.[1]

$$-E_x^{inc}(\vec{r}) = \left[ -j\omega A_x(\vec{r}) - \frac{\partial}{\partial x}\Phi(\vec{r}) \right] \tag{6.1}$$

$$-E_y^{inc}(\vec{r}) = \left[ -j\omega A_y(\vec{r}) - \frac{\partial}{\partial y}\Phi(\vec{r}) \right] \tag{6.2}$$

The magnetic vector potential and electric scalar potential are given by (6.3) and (6.4). The current and charge is confined to the plate; therefore, the integrations are over the plate surface.

$$\vec{A}(\vec{r}) = \frac{\mu}{4\pi} \int \int J(\vec{r}) \frac{e^{-jk|r-\acute{r}|}}{|r-\acute{r}|} \, d\acute{x} \, d\acute{y} \tag{6.3}$$

$$\Phi(\vec{r}) = \frac{1}{4\pi\epsilon} \int \int \rho(\vec{r}) \frac{e^{-jk|r-\acute{r}|}}{|r-\acute{r}|} \, d\acute{x} \, d\acute{y} \tag{6.4}$$

The charge and current are related through the continuity equation:

$$-j\omega\rho = \nabla \cdot \vec{J} = \frac{\partial J_x}{\partial x} + \frac{\partial J_y}{\partial y} \tag{6.5}$$

To implement the moment method, we first expand the $x$ and $y$ components of the surface current ($J_x$ and $J_y$) on the plate, along with the charge ($\rho$). The expansion scheme for current and charge is seen in Figures 6.2, 6.3, 6.4 and 6.5.

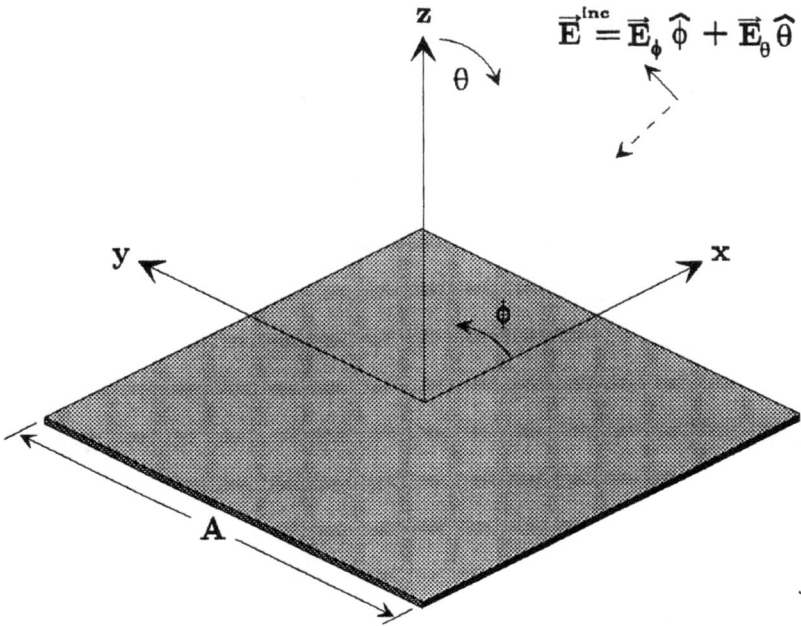

**Figure 6.1** A thin, perfectly conducting metal plate.

$$J_x(\vec{r}) = \sum_{n=1}^{N+1} \sum_{m=1}^{M} J_x^{mn} P_{Jx}^{mn}(\vec{r}) \tag{6.6}$$

$$P_{Jx}^{mn}(\vec{r}) = \begin{cases} 1 & \begin{cases} x_{m-1/2} < x < x_{m+1/2} \\ y_{n-1} < y < y_n \end{cases} \\ 0 & \text{elsewhere} \end{cases} \tag{6.7}$$

$$J_y(\vec{r}) = \sum_{n=1}^{N} \sum_{m=1}^{M+1} J_y^{mn} P_{Jy}^{mn}(\vec{r}) \tag{6.8}$$

$$P_{Jy}^{mn}(\vec{r}) = \begin{cases} 1 & \begin{cases} x_{m-1} < x < x_m \\ y_{n-1/2} < y < y_{n+1/2} \end{cases} \\ 0 & \text{elsewhere} \end{cases} \tag{6.9}$$

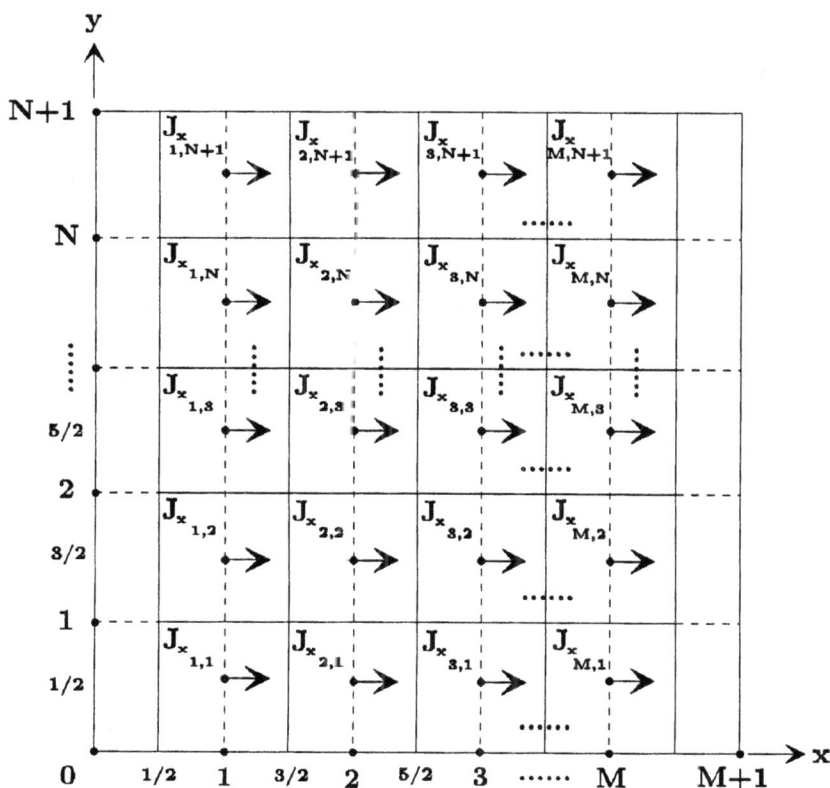

**Figure 6.2** $\vec{J}_x$ current expansion and charge expansion.

We can substitute the current expansions into the magnetic vector potential terms of (6.1) and (6.2). This generates (6.10) and (6.11):

$$A_x(\vec{r}) = \sum_{n=1}^{N+1} \sum_{m=1}^{M} \int_{x_{m-1/2}}^{x_{m+1/2}} \int_{y_{n-1}}^{y_n} J_x^{mn} G(\vec{r}, \vec{r}\,') \, d\acute{y} \, d\acute{x} \tag{6.10}$$

$$A_y(\vec{r}) = \sum_{n=1}^{N} \sum_{m=1}^{M+1} \int_{x_{m-1}}^{x_m} \int_{y_{n-1/2}}^{y_{n+1/2}} J_y^{mn} G(\vec{r}, \vec{r}\,') \, d\acute{y} \, d\acute{x} \tag{6.11}$$

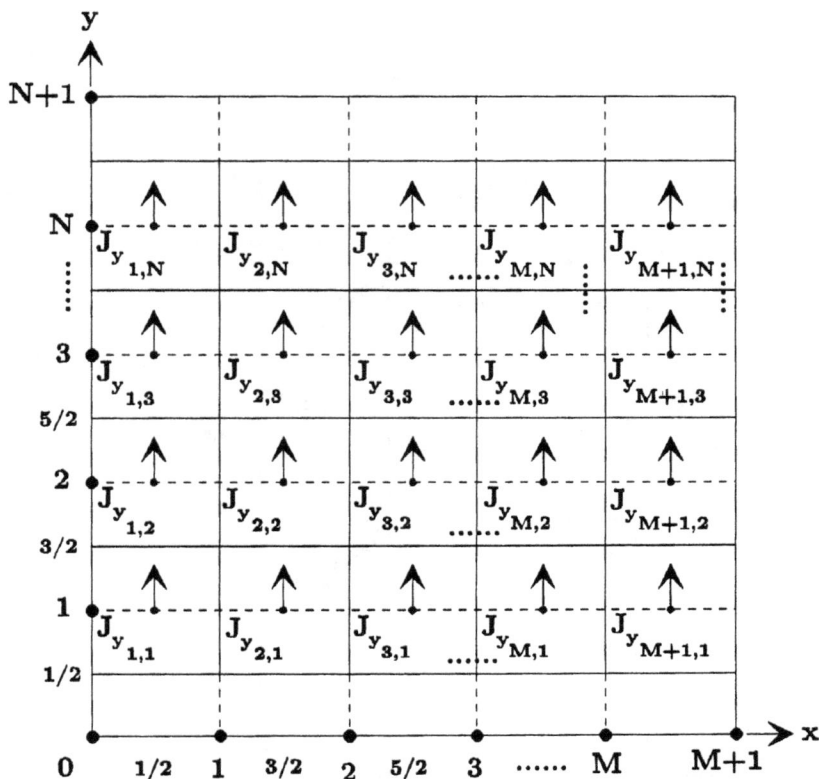

**Figure 6.3** $\vec{J}_y$ current expansion and charge expansion.

Where the free–space Green's function is:

$$G(\vec{r}, \vec{r}) = \frac{e^{-jk|r-\acute{r}|}}{|r-\acute{r}|}$$

The charge expansion functions are given by:

$$P_\rho^{mn}(\vec{r}) = \begin{cases} 1 & \begin{cases} x_{m-1} < x < x_m \\ y_{n-1} < y < y_n \end{cases} \\ 0 \ \text{elsewhere} \end{cases} \tag{6.12}$$

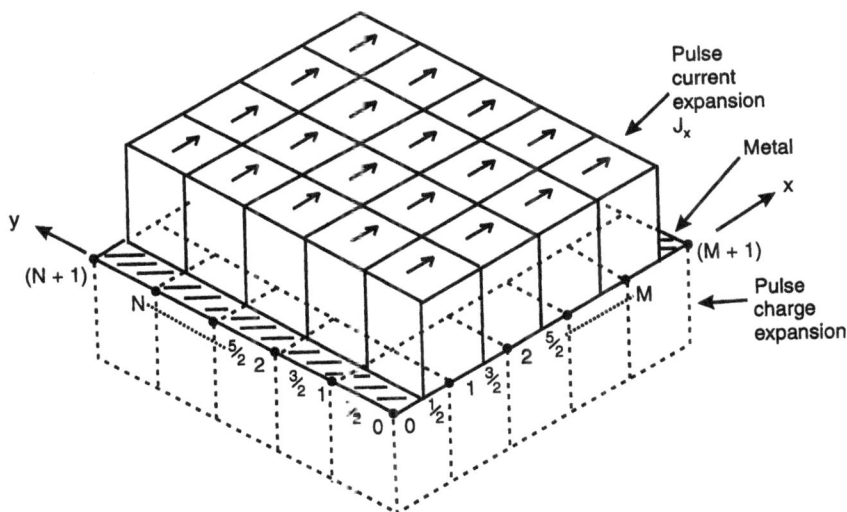

**Figure 6.4**  Three–dimensional view of pulse expansion of current $J_x$ and charge on perfectly conducting plate shown in Figure 6.2.

Finite difference approximations can be introduced into the continuity equation (6.5), which produces:

$$\rho(\vec{r}) = \frac{j}{\omega} \sum_{n=1}^{N+1} \sum_{m=1}^{M+1} \left[ \frac{J_x^{mn} - J_x^{m-1,n}}{\Delta x} + \frac{J_y^{mn} - J_y^{m,n-1}}{\Delta y} \right] P_\rho^{mn}(\vec{r}) \qquad (6.13)$$

The boundary condition on current requires that the normal current at each edge vanish. This allows us to rewrite the summations in (6.13) as:

$$\rho(\vec{r}) = \frac{j}{\omega} \left[ \sum_{n=1}^{N+1} \sum_{m=1}^{M} J_x^{mn} \left( \frac{P_\rho^{mn}(\vec{r})}{\Delta x} - \frac{P_\rho^{m+1,n}(\vec{r})}{\Delta x} \right) + \right.$$

$$\left. \sum_{n=1}^{N} \sum_{m=1}^{M+1} J_y^{mn} \left( \frac{P_\rho^{mn}(r)}{\Delta y} - \frac{P_\rho^{m,n+1}(\vec{r})}{\Delta y} \right) \right] \qquad (6.14)$$

The electric scalar potential terms in (6.1) and (6.2) have a partial derivative acting upon them. For $E_x$ and $E_y$, we may use a weighting (testing) function of

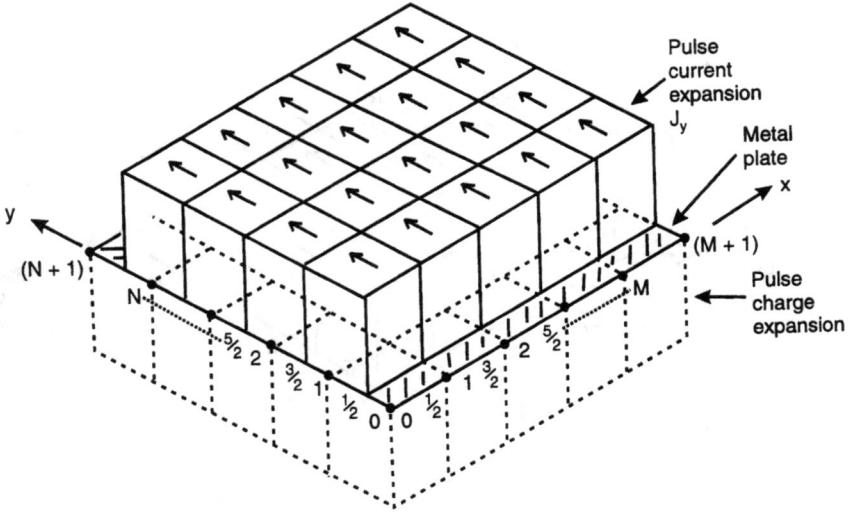

**Figure 6.5**  Three–dimensional view of pulse expansion of current $J_y$, and charge on perfectly conducting plate shown in Figure 6.3.

$P_{J_x}^{mn}(\vec{r})\delta(y - y_{n-1/2})$ and $P_{J_y}^{mn}(\vec{r})\delta(x - x_{m-1/2})$, respectively. The integration with the chosen weighting function removes the partial derivative in (6.1) and (6.2).

For $E_x$, the enforcement function produces:

$$\int_{x_{m-1/2}}^{x_{m+1/2}} -\frac{\partial}{\partial x}\Phi(\vec{r}) \, dx = -\Phi(x_{m+1/2}, y_{n-1/2}) + \Phi(x_{m-1/2}, y_{n-1/2}) \qquad (6.15)$$

For $E_y$, the enforcement function produces:

$$\int_{y_{n-1/2}}^{y_{n+1/2}} -\frac{\partial}{\partial y}\Phi(\vec{r}) \, dy = -\Phi(x_{m-1/2}, y_{n+1/2}) + \Phi(x_{m-1/2}, y_{n-1/2}) \qquad (6.16)$$

The integration in each case removes the partial derivative and the delta functions sift the non–integration variable. Using (6.10) and (6.11), and using (6.15) and (6.16) with (6.1) and (6.2), we obtain:

$$E_x(\vec{r}) = -\sum_{n=1}^{N+1}\sum_{m=1}^{M} j\omega \frac{\mu}{4\tau} \underbrace{\int_{x_{m-1/2}}^{x_{m+1/2}} \int_{y_{n-1}}^{y_n} J_x^{mn} G(\vec{r},\vec{r}\,)\, d\acute{y}\, d\acute{x}}_{A_x}$$

$$-\frac{1}{4\pi\epsilon} \underbrace{\int\int \rho(x_{m+1/2}, y_{n-1/2}, \vec{r}\,) G(x_{m+1/2}, y_{n-1/2}, \acute{r})\, d\acute{y}\, d\acute{x}}_{\Phi(x_{m+1/2}, y_{n-1/2})}$$

$$+\frac{1}{4\pi\epsilon} \underbrace{\int\int \rho(x_{m-1/2}, y_{n-1/2}, \vec{r}\,) G(x_{m-1/2}, y_{n-1/2}, \acute{r})\, d\acute{y}\, d\acute{x}}_{\Phi(x_{m-1/2}, y_{n-1/2})} \qquad (6.17)$$

$$E_y(\vec{r}) = -\sum_{n=1}^{N}\sum_{m=1}^{M+1} j\omega \frac{\mu}{4\pi} \underbrace{\int_{x_{m-1}}^{x_m} \int_{y_{n-1/2}}^{y_{n+1/2}} J_y^{mn} G(\vec{r},\vec{r}\,)\, d\acute{y}\, d\acute{x}}_{A_y}$$

$$-\frac{1}{4\pi\epsilon} \underbrace{\int\int \rho(x_{m-1/2}, y_{n+1/2}, \acute{r}) G(x_{m-1/2}, y_{n+1/2}, \acute{r})\, d\acute{y}\, d\acute{x}}_{\Phi(x_{m-1/2}, y_{n+1/2})}$$

$$+\frac{1}{4\pi\epsilon} \underbrace{\int\int \rho(x_{m-1/2}, y_{n-1/2}, \acute{r}) G(x_{m-1/2}, y_{n-1/2}, \acute{r})\, d\acute{y}\, d\acute{x}}_{\Phi(x_{m-1/2}, y_{n-1/2})} \qquad (6.18)$$

We now insert the expansion of the charge into the scalar potential terms in (6.17) and (6.18). For $E_x$ we obtain:

$$-\frac{1}{4\pi\epsilon}\frac{j}{\omega} \int\int J_x^{mn} \left( \frac{P_\rho^{mn}(\vec{r})}{\Delta x} - \frac{P_\rho^{m+1,n}(\vec{r})}{\Delta x} \right) G(x_{m+1/2}, y_{n-1/2}, \acute{r})\, d\acute{y}\, d\acute{x}$$

$$-\frac{1}{4\pi\epsilon}\frac{j}{\omega} \int\int J_y^{mn} \left( \frac{P_\rho^{mn}(\vec{r})}{\Delta y} - \frac{P_\rho^{m,n+1}(\vec{r})}{\Delta y} \right) G(x_{m+1/2}, y_{n-1/2}, \acute{r})\, d\acute{y}\, d\acute{x}$$

$$+\frac{1}{4\pi\epsilon}\frac{j}{\omega} \int\int J_x^{mn} \left( \frac{P_\rho^{mn}(\vec{r})}{\Delta x} - \frac{P_\rho^{m+1,n}(\vec{r})}{\Delta x} \right) G(x_{m-1/2}, y_{n-1/2}, \acute{r})\, d\acute{y}\, d\acute{x}$$

$$+\frac{1}{4\pi\epsilon}\frac{j}{\omega} \int\int J_y^{mn} \left( \frac{P_\rho^{mn}(\vec{r})}{\Delta y} - \frac{P_\rho^{m,n+1}(\vec{r})}{\Delta y} \right) G(x_{m-1/2}, y_{n-1/2}, \acute{r})\, d\acute{y}\, d\acute{x} \qquad (6.19)$$

For $E_y$

$$-\frac{1}{4\pi\epsilon}\frac{j}{\omega}\int\int J_x^{mn}\left(\frac{P_\rho^{mn}(\vec{r})}{\Delta x}-\frac{P_\rho^{m+1,n}(\vec{r})}{\Delta x}\right)$$

$$G(x_{m-1/2},y_{n+1/2},\vec{r})\,d\acute{y}\,d\acute{x}$$

$$-\frac{1}{4\pi\epsilon}\frac{j}{\omega}\int\int J_y^{mn}\left(\frac{P_\rho^{mn}(\vec{r})}{\Delta y}-\frac{P_\rho^{m,n+1}(\vec{r})}{\Delta y}\right)$$

$$G(x_{m-1/2},y_{n+1/2},\vec{r})\,d\acute{y}\,d\acute{x}$$

$$+\frac{1}{4\pi\epsilon}\frac{j}{\omega}\int\int J_x^{mn}\left(\frac{P_\rho^{mn}(\vec{r})}{\Delta x}-\frac{P_\rho^{m+1,n}(\vec{r})}{\Delta x}\right)$$

$$G(x_{m-1/2},y_{n-1/2},\vec{r})\,d\acute{y}\,d\acute{x}$$

$$+\frac{1}{4\pi\epsilon}\frac{j}{\omega}\int\int J_y^{mn}\left(\frac{P_\rho^{mn}(\vec{r})}{\Delta y}-\frac{P_\rho^{m,n+1}(\vec{r})}{\Delta y}\right)$$

$$G(x_{m-1/2},y_{n-1/2},\vec{r})\,d\acute{y}\,d\acute{x} \qquad (6.20)$$

The domain of each pulse function determines the limits of integration.
For the $E_x$ situation:

$$-\frac{1}{4\pi\epsilon}\frac{j}{\omega}\int_{x_{m-1}}^{x_m}\int_{y_{n-1}}^{y_n}\frac{J_x^{mn}}{\Delta x}G(x_{m+1/2},y_{n-1/2},\vec{r})\,d\acute{y}\,d\acute{x}$$

$$+\frac{1}{4\pi\epsilon}\frac{j}{\omega}\int_{x_m}^{x_{m+1}}\int_{y_{n-1}}^{y_n}\frac{J_x^{mn}}{\Delta x}G(x_{m+1/2},y_{n-1/2},\vec{r})\,d\acute{y}\,d\acute{x}$$

$$-\frac{1}{4\pi\epsilon}\frac{j}{\omega}\int_{x_{m-1}}^{x_m}\int_{y_{n-1}}^{y_n}\frac{J_y^{mn}}{\Delta y}G(x_{m+1/2},y_{n-1/2},\vec{r})\,d\acute{y}\,d\acute{x}$$

$$+\frac{1}{4\pi\epsilon}\frac{j}{\omega}\int_{x_{m-1}}^{x_m}\int_{y_n}^{y_{n+1}}\frac{J_y^{mn}}{\Delta y}G(x_{m+1/2},y_{n-1/2},\vec{r})\,d\acute{y}\,d\acute{x}$$

$$+\frac{1}{4\pi\epsilon}\frac{j}{\omega}\int_{x_{m-1}}^{x_m}\int_{y_{n-1}}^{y_n}\frac{J_x^{mn}}{\Delta x}G(x_{m-1/2},y_{n-1/2},\vec{r})\,d\acute{y}\,d\acute{x}$$

$$-\frac{1}{4\pi\epsilon}\frac{j}{\omega}\int_{x_m}^{x_{m+1}}\int_{y_{n-1}}^{y_n}\frac{J_x^{mn}}{\Delta x}G(x_{m-1/2},y_{n-1/2},\vec{r})\,d\acute{y}\,d\acute{x}$$

$$+\frac{1}{4\pi\epsilon}\frac{j}{\omega}\int_{x_{m-1}}^{x_m}\int_{y_{n-1}}^{y_n}\frac{J_y^{mn}}{\Delta y}G(x_{m-1/2},y_{n-1/2},\vec{r})\,d\acute{y}\,d\acute{x}$$

$$-\frac{1}{4\pi\epsilon}\frac{j}{\omega}\int_{x_{m-1}}^{x_m}\int_{y_n}^{y_{n+1}}\frac{J_y^{mn}}{\Delta y}G(x_{m-1/2},y_{n-1/2},\vec{r})\,d\acute{y}\,d\acute{x} \qquad (6.21)$$

For the $E_y$ situation:

$$-\frac{1}{4\pi\epsilon}\frac{j}{\omega}\int_{x_{m-1}}^{x_m}\int_{y_{n-1}}^{y_n}\frac{J_x^{mn}}{\Delta x}G(x_{m-1/2},y_{n+1/2},\vec{r})\,d\acute{y}\,d\acute{x}$$

$$+\frac{1}{4\pi\epsilon}\frac{j}{\omega}\int_{x_m}^{x_{m-1}}\int_{y_{n-1}}^{y_n}\frac{J_x^{mn}}{\Delta x}G(x_{m-1/2},y_{n+1/2},\vec{r})\,d\acute{y}\,d\acute{x}$$

$$-\frac{1}{4\pi\epsilon}\frac{j}{\omega}\int_{x_{m-1}}^{x_m}\int_{y_{n-1}}^{y_n}\frac{J_y^{mn}}{\Delta y}G(x_{m-1/2},y_{n+1/2},\vec{r})\,d\acute{y}\,d\acute{x}$$

$$+\frac{1}{4\pi\epsilon}\frac{j}{\omega}\int_{x_{m-1}}^{x_m}\int_{y_n}^{y_{n+1}}\frac{J_y^{mn}}{\Delta y}G(x_{m-1/2},y_{n+1/2},\vec{r})\,d\acute{y}\,d\acute{x}$$

$$+\frac{1}{4\pi\epsilon}\frac{j}{\omega}\int_{x_{m-1}}^{x_m}\int_{y_{n-1}}^{y_n}\frac{J_x^{mn}}{\Delta x}G(x_{m-1/2},y_{n-1/2},\vec{r})\,d\acute{y}\,d\acute{x}$$

$$-\frac{1}{4\pi\epsilon}\frac{j}{\omega}\int_{x_m}^{x_{m+1}}\int_{y_{n-1}}^{y_n}\frac{J_x^{mn}}{\Delta x}G(x_{m-1/2},y_{n-1/2},\vec{r})\,d\acute{y}\,d\acute{x}$$

$$+\frac{1}{4\pi\epsilon}\frac{j}{\omega}\int_{x_{m-1}}^{x_m}\int_{y_{n-1}}^{y_n}\frac{J_y^{mn}}{\Delta y}G(x_{m-1/2},y_{n-1/2},\vec{r})\,d\acute{y}\,d\acute{x}$$

$$-\frac{1}{4\pi\epsilon}\frac{j}{\omega}\int_{x_{m-1}}^{x_m}\int_{y_n}^{y_{n+1}}\frac{J_y^{mn}}{\Delta y}G(x_{m-1/2},y_{n-1/2},\vec{r})\,d\acute{y}\,d\acute{x} \qquad (6.22)$$

We can factor the constants and unknown current coefficients in the electric scalar potential terms. When this is done, the $E_x$ case produces:

$$\frac{1}{4\pi\epsilon}\frac{j}{\omega}\frac{J_x^{mn}}{\Delta x}\left[-\int_{x_{m-1}}^{x_m}\int_{y_{n-1}}^{y_n}G(x_{m+1/2},y_{n-1/2},\vec{r})\,d\acute{y}\,d\acute{x}\right.$$

$$+\int_{x_m}^{x_{m+1}}\int_{y_{n-1}}^{y_n}G(x_{m+1/2},y_{n-1/2},\vec{r})\,d\acute{y}\,d\acute{x}$$

$$+\int_{x_{m-1}}^{x_m}\int_{y_{n-1}}^{y_n}G(x_{m-1/2},y_{n-1/2},\vec{r})\,d\acute{y}\,d\acute{x}$$

$$-\int_{x_m}^{x_{m+1}}\int_{y_{n-1}}^{y_n} G(x_{m-1/2}, y_{n-1/2}, \vec{r})\, d\acute{y}\, d\acute{x} \Bigg]$$

$$+\frac{1}{4\pi\epsilon}\frac{j}{\omega}\frac{J_y^{mn}}{\Delta y}\Bigg[-\int_{x_{m-1}}^{x_m}\int_{y_{n-1}}^{y_n} G(x_{m+1/2}, y_{n-1/2}, \vec{r})\, d\acute{y}\, d\acute{x} \qquad (6.23)$$

$$+\int_{x_{m-1}}^{x_m}\int_{y_n}^{y_{n+1}} G(x_{m+1/2}, y_{n-1/2}, \vec{r})\, d\acute{y}\, d\acute{x}$$

$$+\int_{x_{m-1}}^{x_m}\int_{y_{n-1}}^{y_n} G(x_{m-1/2}, y_{n-1/2}, \vec{r})\, d\acute{y}\, d\acute{x}$$

$$-\int_{x_{m-1}}^{x_m}\int_{y_n}^{y_{n+1}} G(x_{m-1/2}, y_{n-1/2}, \vec{r})\, d\acute{y}\, d\acute{x} \Bigg] \qquad (6.24)$$

For $Ey$

$$\frac{1}{4\pi\epsilon}\frac{j}{\omega}\frac{J_x^{mn}}{\Delta x}\Bigg[-\int_{x_{m-1}}^{x_m}\int_{y_{n-1}}^{y_n} G(x_{m-1/2}, y_{n+1/2}, \vec{r})\, d\acute{y}\, d\acute{x}$$

$$+\int_{x_m}^{x_{m+1}}\int_{y_{n-1}}^{y_n} G(x_{m-1/2}, y_{n+1/2}, \vec{r})\, d\acute{y}\, d\acute{x}$$

$$+\int_{x_{m-1}}^{x_m}\int_{y_{n-1}}^{y_n} G(x_{m-1/2}, y_{n-1/2}, \vec{r})\, d\acute{y}\, d\acute{x}$$

$$-\int_{x_m}^{x_{m+1}}\int_{y_{n-1}}^{y_n} G(x_{m-1/2}, y_{n-1/2}, \vec{r})\, d\acute{y}\, d\acute{x} \Bigg] \qquad (6.25)$$

$$+\frac{1}{4\pi\epsilon}\frac{j}{\omega}\frac{J_y^{mn}}{\Delta y}\Bigg[-\int_{x_{m-1}}^{x_m}\int_{y_{n-1}}^{y_n} G(x_{m-1/2}, y_{n+1/2}, \vec{r})\, d\acute{y}\, d\acute{x}$$

$$+\int_{x_{m-1}}^{x_m}\int_{y_n}^{y_{n+1}} G(x_{m-1/2}, y_{n+1/2}, \vec{r})\, d\acute{y}\, d\acute{x}$$

$$+\int_{x_{m-1}}^{x_m}\int_{y_{n-1}}^{y_n} G(x_{m-1/2}, y_{n-1/2}, \vec{r})\, d\acute{y}\, d\acute{x}$$

$$-\int_{x_{m-1}}^{x_m}\int_{y_n}^{y_{n+1}} G(x_{m-1/2}, y_{n-1/2}, \vec{r})\, d\acute{y}\, d\acute{x} \Bigg] \qquad (6.26)$$

We have not chosen an enforcement (weighting, testing) function for the magnetic vector potential. We choose the same pulse function we used for the scalar case. The integration may be implemented with the simple approximation of (6.27):

$$\int_{\Delta x} f(x) \, dx \approx f(\text{midpoint}) \cdot \Delta x \tag{6.27}$$

The use of this approximation produces:

$$A_x \approx \int_{x_{m-1/2}}^{x_{m+1/2}} \int_{y_{n-1}}^{y_n} J_x^{mn} G(x_m, y_{n-1/2}, \vec{r}) \, d\acute{y} \, d\acute{x} \cdot \Delta x_m \tag{6.28}$$

$$A_y \approx \int_{x_{m-1}}^{x_m} \int_{y_{n-1/2}}^{y_{n+1/2}} J_y^{mn} G(x_{m-1/2}, y_n, \vec{r}) \, d\acute{y} \, d\acute{x} \cdot \Delta y_n \tag{6.29}$$

For the incident electric field:

$$\int \vec{E}_x \, dx \approx E_{x_i, y_{j-1/2}} \cdot \Delta x = E_x^{mn} \tag{6.30}$$

$$\int \vec{E}_y \, dy \approx E_{x_{i-1/2}, y_j} \cdot \Delta y = E_y^{mn} \tag{6.31}$$

We can abbreviate the equations we developed as:

$$E_x^{ij} = \sum_{n=1}^{N+1} \sum_{m=1}^{M} J_x^{mn} \left( A_x^{mn} + \Phi_x^{mn} \right) + \sum_{n=1}^{N} \sum_{m=1}^{M+1} J_y^{mn} \Phi_y^{mn} \tag{6.32}$$

$$E_y^{ij} = \sum_{n=1}^{N+1} \sum_{m=1}^{M} J_x^{mn} \Phi_x^{mn} + \sum_{n=1}^{N} \sum_{m=1}^{M+1} J_y^{mn} \left( A_y^{mn} + \Phi_y^{mn} \right) \tag{6.33}$$

If $M = N$ the first equation has $M(N+1) = N(M+1)$ equations in $2M(N+1)$ or $2N(M+1)$ unknowns. The second equation provides the required equations to produce a solution. Substituting we obtain:

$$E_x^{ij} =$$

$$\sum_{n=1}^{N+1} \sum_{m=1}^{M} J_x^{mn} \left[ -j\omega \frac{\mu}{4\pi} \int_{x_{m-1/2}}^{x_{m-1/2}} \int_{y_{n-1}}^{y_n} G(x_i, y_{j-1/2}, \vec{r}) \, d\acute{y} \, d\acute{x} \cdot \Delta x_i \right.$$

$$+ \frac{j}{4\pi\omega\epsilon\Delta x} \left( - \int_{x_{m-1}}^{x_m} \int_{y_{n-1}}^{y_n} G(x_{i+1/2}, y_{j-1/2}, \vec{r}) \, d\acute{y} \, d\acute{x} \right.$$

$$+ \int_{x_m}^{x_{m+1}} \int_{y_{n-1}}^{y_n} G(x_{i+1/2}, y_{j-1/2}, \vec{r}) \, d\acute{y} \, d\acute{x}$$

$$+ \int_{x_{n-1}}^{x_m} \int_{y_{n-1}}^{y_n} G(x_{i-1/2}, y_{j-1/2}, \vec{r}) \, d\acute{y} \, d\acute{x}$$

$$- \int_{x_m}^{x_{m+1}} \int_{y_{n-1}}^{y_n} G(x_{i-1/2}, y_{j-1/2}, \vec{r}) \, d\acute{y} \, d\acute{x} \Bigg]$$

$$+ \sum_{n=1}^{N} \sum_{m=1}^{M+1} J_y^{mn} \frac{j}{4\pi\omega\epsilon\Delta y} \Bigg[ - \int_{x_{m-1}}^{x_m} \int_{y_{n-1}}^{y_n} G(x_{i+1/2}, y_{j-1/2}, \vec{r}) \, d\acute{y} \, d\acute{x}$$

$$+ \int_{x_{m-1}}^{x_m} \int_{y_n}^{y_{n+1}} G(x_{i+1/2}, y_{j-1/2}, \vec{r}) \, d\acute{y} \, d\acute{x}$$

$$+ \int_{x_{m-1}}^{x_m} \int_{y_{n-1}}^{y_n} G(x_{i-1/2}, y_{j-1/2}, \vec{r}) \, d\acute{y} \, d\acute{x}$$

$$- \int_{x_{m-1}}^{x_m} \int_{y_n}^{y_{n+1}} G(x_{i-1/2}, y_{j-1/2}, \vec{r}) \, d\acute{y} \, d\acute{x} \Bigg] \tag{6.34}$$

$$E_y^{ij} = \sum_{n=1}^{N+1} \sum_{m=1}^{M} J_x^{mn} \frac{j}{4\pi\omega\epsilon\Delta x} \Bigg[ - \int_{x_{m-1}}^{x_m} \int_{y_{n-1}}^{y_n} G(x_{i-1/2}, y_{j+1/2}, \vec{r}) \, d\acute{y} \, d\acute{x}$$

$$+ \int_{x_m}^{x_{m+1}} \int_{y_{n-1}}^{y_n} G(x_{i-1/2}, y_{j+1/2}, \vec{r}) \, d\acute{y} \, d\acute{x}$$

$$+ \int_{x_{m-1}}^{x_m} \int_{y_{n-1}}^{y_n} G(x_{i-1/2}, y_{j-1/2}, \vec{r}) \, d\acute{y} \, d\acute{x}$$

$$- \int_{x_m}^{x_{m+1}} \int_{y_{n-1}}^{y_n} G(x_{i-1/2}, y_{j-1/2}, \vec{r}) \, d\acute{y} \, d\acute{x} \Bigg]$$

$$\sum_{n=1}^{N} \sum_{m=1}^{M+1} J_y^{mn} \Bigg[ -j\omega \frac{\mu}{4\pi} \int_{x_{m-1}}^{x_m} \int_{y_{n-1/2}}^{y_{n+1/2}} G(x_{i-1/2}, y_j, \vec{r}) \, d\acute{y} \, d\acute{x} \cdot \Delta y_j$$

$$+ \frac{j}{4\pi\omega\epsilon\Delta y} \Bigg( - \int_{x_{m-1}}^{x_m} \int_{y_{n-1}}^{y_n} G(x_{i-1/2}, y_{j+1/2}, \vec{r}) \, d\acute{y} \, d\acute{x}$$

$$+ \int_{x_{m-1}}^{x_m} \int_{y_n}^{y_{n+1}} G(x_{i-1/2}, y_{j+1/2}, \vec{r}) \, d\acute{y} \, d\acute{x}$$

$$+ \int_{x_{m-1}}^{x_m} \int_{y_{n-1}}^{y_n} G(x_{i-1/2}, y_{j-1/2}, \vec{r}) \, d\acute{y} \, d\acute{x}$$

$$- \int_{x_{m-1}}^{x_m} \int_{y_n}^{y_{n+1}} G(x_{i-1/2}, y_{j-1/2}, \vec{r}) \, d\acute{y} \, d\acute{x} \Bigg) \Bigg] \tag{6.35}$$

The electric field of the incoming plane wave is described by:

$$E_x^{inc}(x,y)\hat{x} + E_y^{inc}(x,y)\hat{y} =$$

$$[(E_\theta \cos\theta \cos\phi - E_\phi \sin\phi)\,\hat{x}$$

$$+ (E_\theta \cos\theta \sin\phi + E_\phi \cos\phi)\,\hat{y}]\,e^{jk(x\sin\theta\cos\phi + y\sin\theta\sin\phi)} \qquad (6.36)$$

We match the electric field values at the center of each current expansion function. This produces:

$$E_x^{ij} = [E_\theta \cos\theta \cos\phi - E_\phi \sin\phi]\,e^{jk\left(x_i \sin\theta\cos\phi + y_{j-1/2}\sin\theta\sin\phi\right)} \qquad (6.37)$$

$$E_y^{ij} = [E_\theta \cos\theta \sin\phi + E_\phi \cos\phi]\,e^{jk\left(x_{i-1/2}\sin\theta\cos\phi + y_j\sin\theta\sin\phi\right)} \qquad (6.38)$$

We choose our observation points independently (the $i, j$ indices) to obtain as many equations as unknowns. This allows one to calculate the current components $J_x$ and $J_y$. This has been done for a $\lambda/2$ plate. The current components are from a plane wave at normal incidence. The incoming plane wave is $\hat{\theta}$ polarized (i.e., along the $x$-axis).

The magnitude of the current components, $J_x$ and $J_y$, calculated using (6.34), (6.35), (6.37), and (6.38) are presented in Figure 6.6. The plane wave is at normal incidence to the plate. We note the one over square root singularity for the $\vec{J}_x$ current component. This is similar to what we saw in the two-dimensional TM strip solution. The cross–polarized current rises at the corners and has a much lower level as expected. This is also consistent with the two-dimensional TE strip solution.

This current shows the characteristics we would expect from our calculations of the current generated on infinitely long strips in Chapter 4. In the TM case for an infinite strip, we note that current running along the edge of a strip increases in a singular manner (Figure 4.5). In the TE case of an infinite strip, it is seen that current normal to an edge goes to zero (Figure 4.6). We see both situations when we view Figure 6.6. The current component with the largest magnitude is the $J_x$ component. It exhibits the properties expected from the study of infinite conducting strips. The $J_y$ component is the current generated that is not along the incoming waves polarization. We note that it is much smaller in magnitude.

We can now calculate the RCS from these currents. RCS is defined as:

$$\sigma = \lim_{r\to\infty} 4\pi r^2 \frac{|E^s|^2}{|E^{inc}|^2} \qquad (6.39)$$

The $x$ and $y$ components of the scattered electric field are given by:

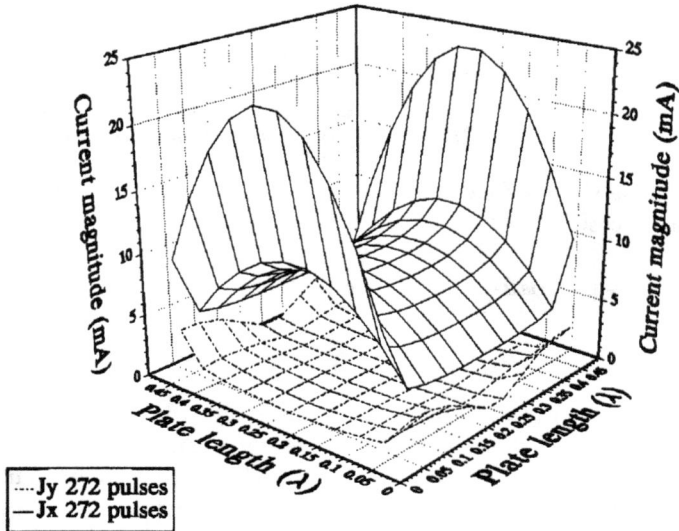

**Figure 6.6** The current $\mid \vec{J}_x \mid$ and $\mid \vec{J}_y \mid$ on a $\lambda/2$ plate calculated using (6.34) and (6.35) at incident angle of $\phi = 0$ and $\theta = 0$ with $\mid \vec{E}_\theta \mid = 1$ V/m $\mid \vec{E}_\phi \mid = 0$ V/m.

$$E_x^s = \frac{-j\omega\mu}{4\pi}\frac{e^{-jkr}}{r}\int\int J_x(\acute{x},\acute{y})e^{jk(\acute{x}\sin\theta\cos\phi+\acute{y}\sin\theta\sin\phi)}d\acute{s} \qquad (6.40)$$

$$E_y^s = \frac{-j\omega\mu}{4\pi}\frac{e^{-jkr}}{r}\int\int J_y(\acute{x},\acute{y})e^{jk(\acute{x}\sin\theta\cos\phi+\acute{y}\sin\theta\sin\phi)}d\acute{s} \qquad (6.41)$$

The incoming electric field is defined in terms of $E_\theta$ and $E_\phi$. Converting, we obtain:

$$E_\theta^s = E_x\cos\theta\cos\phi + E_y\cos\theta\sin\phi \qquad (6.42)$$

$$E_\phi^s = -E_x\sin\phi + E_y\cos\phi \qquad (6.43)$$

We define functions for the integrations in (6.40) and (6.41):

$$I_x(\theta, \phi) = \int \int J_x(\acute{x}, \acute{y}) e^{jk(\acute{x} \sin \theta \cos \phi + \acute{y} \sin \theta \sin \phi)} d\acute{s} \tag{6.44}$$

$$I_y(\theta, \phi) = \int \int J_y(\acute{x}, \acute{y}) e^{jk(\acute{x} \sin \theta \cos \phi + \acute{y} \sin \theta \sin \phi)} d\acute{s} \tag{6.45}$$

Using (6.39), we obtain:

$$\sigma_\theta = \frac{\omega^2 \mu^2}{4\pi} \frac{\mid I_x(\theta, \phi) \cos \theta \cos \phi + I_y(\theta, \phi) \cos \theta \sin \phi \mid^2}{\mid E_\theta^{inc} \mid^2} \tag{6.46}$$

$$\sigma_\phi = \frac{\omega^2 \mu^2}{4\pi} \frac{\mid -I_x(\theta, \phi) \sin \phi + I_y(\theta, \phi) \cos \phi \mid^2}{\mid E_\phi^{inc} \mid^2} \tag{6.47}$$

The integrations in (6.44) and (6.45) may be approximated with:

$$I_x(\theta, \phi) \approx \Delta s \sum_{m=1}^{M} \sum_{n=1}^{N+1} J_x^{m,n} e^{jk(x_m \sin \theta \cos \phi + y_{n-1/2} \sin \theta \sin \phi)} \tag{6.48}$$

$$I_y(\theta, \phi) \approx \Delta s \sum_{m=1}^{M+1} \sum_{n=1}^{N} J_y^{m,n} e^{jk(x_{m-1/2} \sin \theta \cos \phi + y_n \sin \theta \sin \phi)} \tag{6.49}$$

### 6.1.2  Moment Method Solution (Rooftop/Pulse)

Thus far, we expanded the current and charge as two–dimensional pulse functions. In Chapter 3 we expanded the current on a wire using overlapping triangle functions. We can use a similar expansion to expand the current on a two–dimensional surface (Figures 6.7 and 6.8). The function is triangular along one axis and constant on the other. These functions are often called *rooftop* functions because of their resemblance to the roof of a house. If we expand the current in this manner, we discover that only sight modifications in our prior formulation are required to implement rooftop expansion functions. Equations (6.6), (6.7), (6.8), and (6.9) will become:

$$J_x(\vec{r}) = \sum_{n=1}^{N+1} \sum_{m=1}^{M} J_x^{mn} T_{J_x}^{mn}(x - x_m) \tag{6.50}$$

$$T_{J_x}^{mn}(\vec{r}) = \begin{cases} 1 - \frac{\mid x \mid}{H} & \begin{cases} x_{m-1} < x < x_{m+1} \\ y_{n-1} < y < y_n \end{cases} \\ 0 \quad \text{elsewhere} \end{cases} \tag{6.51}$$

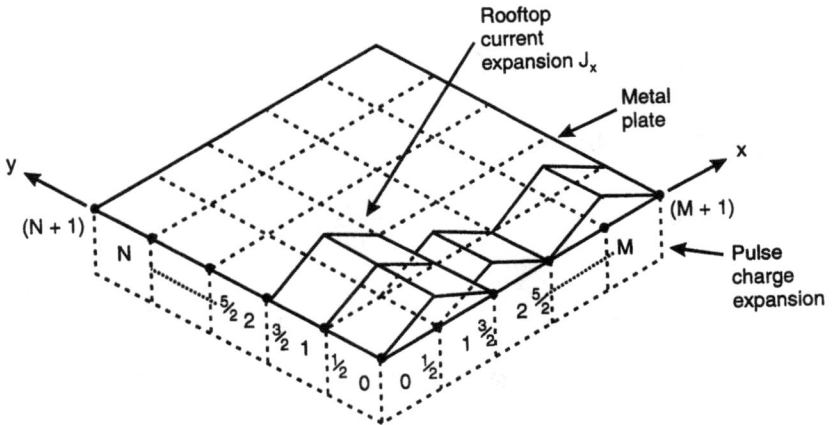

**Figure 6.7** Three–dimensional view of rooftop expansion of current $J_x$ and pulse expansion of charge on perfectly conducting plate.

$$J_y(\vec{r}) = \sum_{n=1}^{N} \sum_{m=1}^{M+1} J_y^{mn} T_{J_y}^{mn}(y - y_n) \qquad (6.52)$$

$$T_{J_y}^{mn}(\vec{r}) = \begin{cases} 1 - \dfrac{|y|}{H} & \begin{cases} x_{m-1} < x < x_m \\ y_{n-1} < y < y_{n+1} \end{cases} \\ 0 \quad \text{elsewhere} \end{cases} \qquad (6.53)$$

$$A_x \approx \int_{x_{m-1}}^{x_{m+1}} \int_{y_{n-1}}^{y_n} J_x^{mn} G(x_m, y_{n-1/2}, \vec{r}) T_{J_x}^{mn}(\acute{x} - x_m) \, d\acute{y} \, d\acute{x} \cdot \Delta x_m \qquad (6.54)$$

$$A_y \approx \int_{x_{m-1}}^{x_m} \int_{y_{n-1}}^{y_{n+1}} J_y^{mn} G(x_{m-1/2}, y_n, \vec{r}) T_{J_y}^{mn}(\acute{y} - y_n) \, d\acute{y} \, d\acute{x} \cdot \Delta y_n \qquad (6.55)$$

Equations (6.34) and (6.35) become:

$$E_x^{mn} =$$

$$\sum_{n=1}^{N+1} \sum_{m=1}^{M} J_x^{mn} \left[ -j\omega \frac{\mu}{4\pi} \int_{x_{m-1}}^{x_{m+1}} \int_{y_{n-1}}^{y_n} J_x^{mn} G(x_i, y_{j-1/2}, \vec{r}) \right.$$

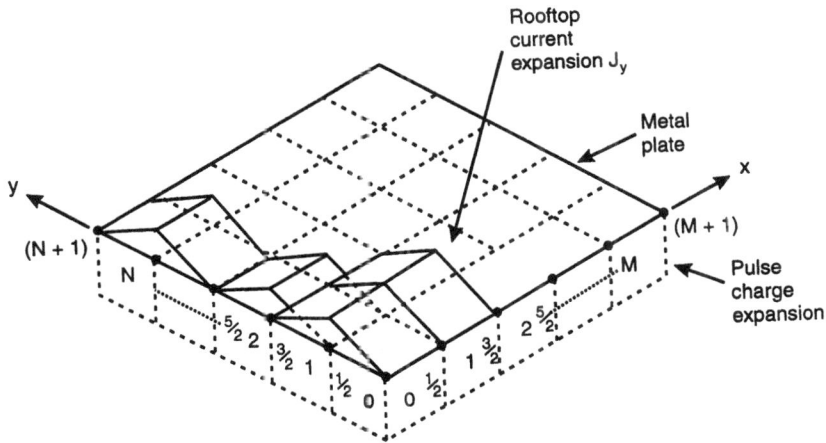

**Figure 6.8**  Three–dimensional view of rooftop expansion of current $J_y$ and pulse expansion of charge on perfectly conducting plate.

$$T_{J_x}^{mn}(\acute{x} - x_m)\, d\acute{y}\, d\acute{x} \cdot \Delta x_i$$

$$+\frac{j}{4\pi\omega\epsilon\Delta x}\left(-\int_{x_{m-1}}^{x_m}\int_{y_{n-1}}^{y_n} G(x_{i+1/2}, y_{j-1/2}, \vec{r})\, d\acute{y}\, d\acute{x}\right.$$

$$+\int_{x_m}^{x_{m+1}}\int_{y_{n-1}}^{y_n} G(x_{i+1/2}, y_{j-1/2}, \vec{r})\, d\acute{y}\, d\acute{x}$$

$$+\int_{x_{m-1}}^{x_m}\int_{y_{n-1}}^{y_n} G(x_{i-1/2}, y_{j-1/2}, \vec{r})\, d\acute{y}\, d\acute{x}$$

$$\left.-\int_{x_m}^{x_{m+1}}\int_{y_{n-1}}^{y_n} G(x_{i-1/2}, y_{j-1/2}, \vec{r})\, d\acute{y}\, d\acute{x}\right]$$

$$+\sum_{n=1}^{N}\sum_{m=1}^{M+1} J_y^{mn}\frac{j}{4\pi\omega\epsilon\Delta y}\left[-\int_{x_{m-1}}^{x_m}\int_{y_{n-1}}^{y_n} G(x_{i+1/2}, y_{j-1/2}, \vec{r})\, d\acute{y}\, d\acute{x}\right.$$

$$+\int_{x_{m-1}}^{x_m}\int_{y_n}^{y_{n+1}} G(x_{i+1/2}, y_{j-1/2}, \vec{r})\, d\acute{y}\, d\acute{x}$$

$$+ \int_{x_{m-1}}^{x_m} \int_{y_{n-1}}^{y_n} G(x_{i-1/2}, y_{j-1/2}, \vec{r}) \, d\acute{y} \, d\acute{x}$$

$$\left. - \int_{x_{m-1}}^{x_m} \int_{y_n}^{y_{n+1}} G(x_{i-1/2}, y_{j-1/2}, \vec{r}) \, d\acute{y} \, d\acute{x} \right] \tag{6.56}$$

$$E_y^{mn} = \sum_{n=1}^{N+1} \sum_{m=1}^{M} J_x^{mn} \frac{j}{4\pi\omega\epsilon\Delta x} \left[ -\int_{x_{m-1}}^{x_m} \int_{y_{n-1}}^{y_n} G(x_{i-1/2}, y_{j+1/2}, \vec{r}) \, d\acute{y} \, d\acute{x} \right.$$

$$+ \int_{x_m}^{x_{m+1}} \int_{y_{n-1}}^{y_n} G(x_{i-1/2}, y_{j+1/2}, \vec{r}) \, d\acute{y} \, d\acute{x}$$

$$+ \int_{x_{m-1}}^{x_m} \int_{y_{n-1}}^{y_n} G(x_{i-1/2}, y_{j-1/2}, \vec{r}) \, d\acute{y} \, d\acute{x}$$

$$\left. - \int_{x_m}^{x_{m+1}} \int_{y_{n-1}}^{y_n} G(x_{i-1/2}, y_{j-1/2}, \vec{r}) \, d\acute{y} \, d\acute{x} \right]$$

$$\sum_{n=1}^{N} \sum_{m=1}^{M+1} J_y^{mn} \left[ -j\omega\frac{\mu}{4\pi} \int_{x_{m-1}}^{x_m} \int_{y_{n-1}}^{y_{n+1}} J_y^{mn} G(x_{i-1/2}, y_j, \vec{r}) \right.$$

$$T_{J_y}^{mn}(\acute{y} - y_n) \, d\acute{y} \, d\acute{x} \cdot \Delta y_j$$

$$+ \frac{j}{4\pi\omega\epsilon\Delta y} \left( -\int_{x_{m-1}}^{x_m} \int_{y_{n-1}}^{y_n} G(x_{i-1/2}, y_{j+1/2}, \vec{r}) \, d\acute{y} \, d\acute{x} \right.$$

$$+ \int_{x_{m-1}}^{x_m} \int_{y_n}^{y_{n+1}} G(x_{i-1/2}, y_{j+1/2}, \vec{r}) \, d\acute{y} \, d\acute{x}$$

$$+ \int_{x_{m-1}}^{x_m} \int_{y_{n-1}}^{y_n} G(x_{i-1/2}, y_{j-1/2}, \vec{r}) \, d\acute{y} \, d\acute{x}$$

$$\left. \left. - \int_{x_{m-1}}^{x_m} \int_{y_n}^{y_{n+1}} G(x_{i-1/2}, y_{j-1/2}, \vec{r}) \, d\acute{y} \, d\acute{x} \right) \right] \tag{6.57}$$

**Table 6.1**

Radar Cross Section of a One Wavelength Square Plate
Using Pulse Functions

| N | Sigma | Extrapolation |
|---|---|---|
| 2 | 6.51 | 6.51 |
| 4 | 8.23 | 9.96 |
| 8 | 9.83 | 11.91 |
| 16 | 10.66 | 11.47 |

### 6.1.3 Numerical Results

The currents calculated using 240 basis functions of either Pulses or Rooftops are indistinguishable when graphed. A situation where a difference between the numerical solutions is easily observed occurs when we extrapolate the RCS data of a plate. Table 6.1 presents a moment method calculation of the RCS produced by a $1\lambda$ plate using the pulse/pulse formulation with an incoming wave at normal incidence. The moment method solution has not converged and the extrapolation does not become monotonic.

In Table 6.2 we find moment method results for the same $1\lambda$ plate using rooftops for the current, pulses for the charge, and pulse weighting. The calculation values start larger than in the pulse/pulse solution and extrapolate to 11.49 $\pm 0.1$ for the normalized RCS value. In Chapter 3, we saw that using of triangles as basis functions produced a monotonic extrapolation much faster than when pulses are used. It appears this is also true when one uses rooftops, instead of two–dimensional pulse functions, to solve for the RCS of a flat plate. While this appears to be true, it is good to exercise caution when the computational limit (in this case, the size of matrix a typical PC can calculate) leaves us with a result that appears monotonic but does so for only two extrapolation points. If we have three data points that agree in the extrapolation column, we could be more confident it is an accurate answer. It is important to remain cautious when interpreting data that appears to converge. Extrapolated data can sometimes appear to converge and then quickly diverge. An example of this is found in Table 3.2. The solution for the reduced kernel scattering from a wire could appear to converge if we only could compute to

**Table 6.2**

Radar Cross Section of a One Wavelength Square Plate
Using Rooftop Current and Pulse Charge Functions

| N  | Sigma | Extrapolation |
|----|-------|---------------|
| 2  | 7.52  | 7.52          |
| 4  | 8.87  | 10.21         |
| 8  | 10.02 | 11.50         |
| 16 | 10.72 | 11.49         |

$M = 16$. Admittedly, the error is not large, but we know that only after we have calculated an answer in which we have considerable confidence.

The literature available on the RCS of a flat plate provides some confidence in the extrapolated rooftop solution. One recent solution produced a value of approximately 11.25 for the normalized RCS of a $1\lambda$ plate.[2] This value appears to have been calculated with 45 basis functions using a method that separates the surface current into even and odd modes. The convergence of this method appears to be very rapid; but as no convergence data is offered, 11.49 appears just as valid an estimate.

Work done before modern computational capabilities by Rahmat–Samii and Mittra offeres a value of approximately 10.1 for a normalized RCS using a moment method solution.[3] They compare this estimate of RCS to a physical optics solution and a GTD solution.[4] Experimental data offered by Ross is for plates of $3.125\lambda$, $3.91\lambda$, and $5.08\lambda$ per side. These plates are all of a dimension larger than a current PC's capability to provide an accurate moment method solution.

It is interesting to note how current distribution changes as the plate size is increased. A one–wavelength plate has a single half–sinusiodal distribution (one peak) as in Figure 6.9. A two–wavelength plate has a full sinusiod (two peaks), as in Figure 6.10. A three–wavelength plate has a cycle and one–half (three peaks) as in Figure 6.11.

We note that the imaginary current magnitude is larger than that of the real. If we compare the current at the edges of the plate, where it is maximum, it resembles the current seen for a single wire of comparable length. The current on a 1,–2,–and $3\lambda$ long wire is presented in Figures 6.12, 6.13, and 6.14. The imaginary and real currents are very similar for both cases. This illustrates that a reasonable amount

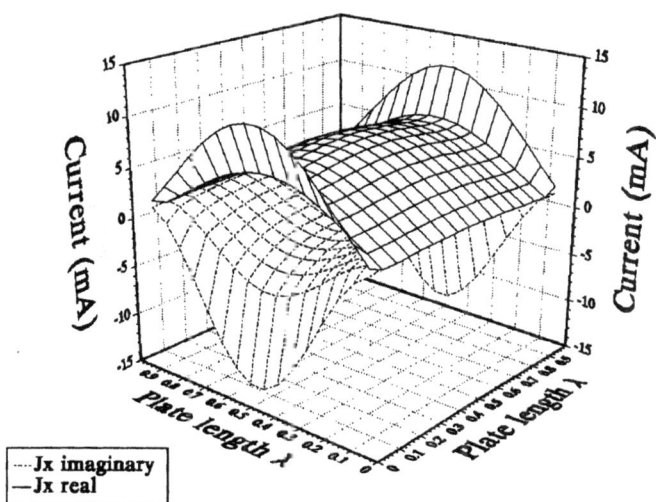

**Figure 6.9** Current on $1\lambda$ plate produced by a electromagnetic plane wave at normal incidence polarized along the $x$-axis.

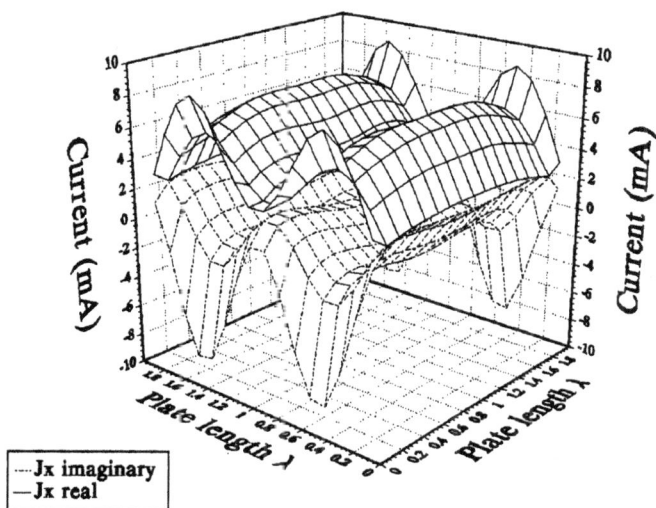

**Figure 6.10** Current on $2\lambda$ plate produced by a electromagnetic plane wave at normal incidence polarized along the $x$-axis.

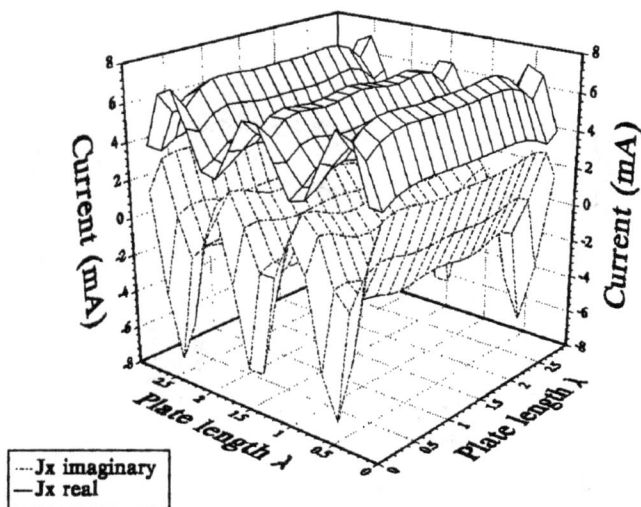

**Figure 6.11**  Current on $3\lambda$ plate produced by a electromagnetic plane wave at normal incidence polarized along the $x$–axis.

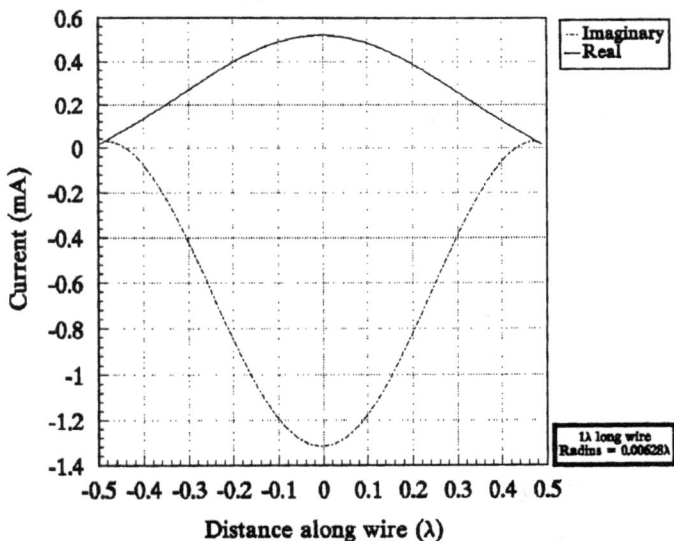

Figure **6.12**  Current on a $1\lambda$ wire of radius $0.00628\lambda$.

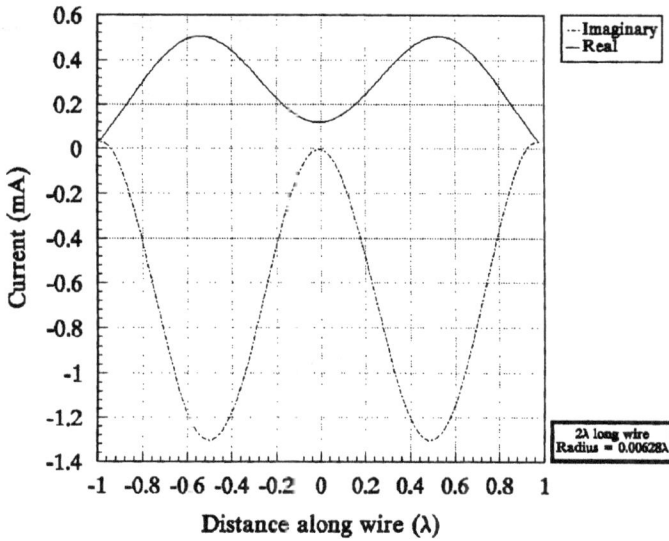

**Figure 6.13** Current on a $2\lambda$ wire of radius $0.00628\lambda$.

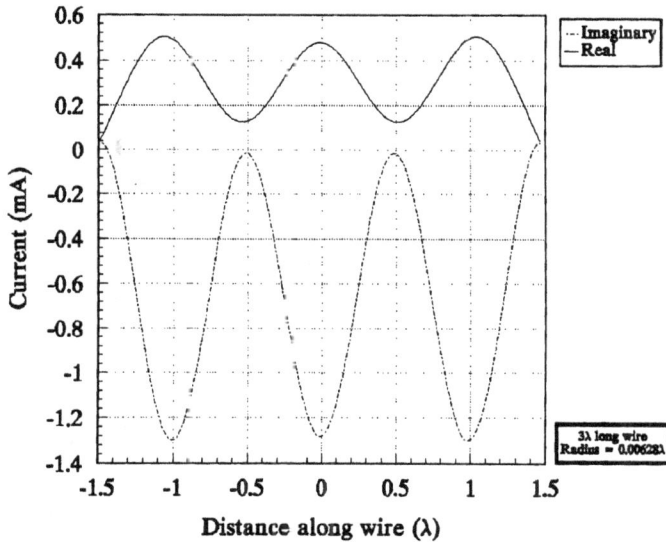

**Figure 6.14** Current on a $3\lambda$ wire of radius $0.00628\lambda$.

of heuristic information about the scattering from a three–dimensional object may be obtained from one, and two–dimensional scattering characteristics.

One might wonder if these heuristic similarities could be translated into estimates. When comparing the measured RCS of a flat plate to a physical optics and GTD solution, Ross introduced an approximate equation which relates the RCS of a two dimensional strip to three dimensional flat plate RCS. A special case of this equation for our square plate is presented as:

$$\sigma(area) = \frac{2A^2}{\lambda^2}\sigma(length) \tag{6.58}$$

The accuracy of this equation may be evaluated using the moment method solution for infinite strips in Chapter 4. The solution for a strip is less computationally intensive than a flat plate calculation. The extrapolations for a strip solution clearly extrapolates in a monotonic fashion. This provides a stable baseline to compare (6.58) with moment method estimates for a flat plate.

### Table 6.3

Comparison of RCS values obtained
using a 2D to 3D transformation and a flat plate solution

| Plate Size | 2D to 3D RCS | Plate RCS |
|------------|--------------|-----------|
| $(A/\lambda)$ | $(\sigma/\lambda^2)$ | $(\sigma/\lambda^2)$ |
| 0.25 | 0.07 | 0.08 |
| 0.50 | 0.84 | 2.18 |
| 0.75 | 4.16 | 5.39 |
| 1.00 | 12.97 | 11.49 |
| 1.25 | 31.14 | 27.18 |

The RCS values predicted by (6.58) are close to the moment method calculations except at 0.5λ. The data appears to extrapolate up to 1.25λ. It appears that the two–dimesional to three–dimensional RCS equation becomes more accurate as the size of a plate increases. As mentioned, Ross presents measured data for three different plate dimensions. In terms of wavelengths, they are beyond the ability of a typical PC to calculate accurately. The two–dimensional moment method solution converges nicely for these sizes. Comparisons of the two–dimensional to

**Table 6.4**

Comparison of RCS values of a

flat plate measurements with 2D to 3D transformation

Source: Ross

$\lambda = 1.28$ inches

| Plate Side Length A | Measured RCS $\sigma$ | 2D to 3D RCS $\sigma$ |
|---|---|---|
| (inches) | (dBsm) | (dBsm) |
| 4.0 | 0.61 ±0.2 | 1.04 |
| 5.0 | 5.15 ±0.3 | 4.91 |
| 6.5 | 9.00 ±0.3 | 9.47 |

three–dimensional predictions to measured data presented by Ross are offered in Table 6.4.

We note that the two–dimensional to three-dimensional predictions are close to the measured data. The differences in RCS between the predictions and data are on the order of tenths of a dB. Tables 6.3 and 6.4 provide evidence that we can obtain a usable amount of quantitative data using (6.58).

## 6.2  Concluding Remarks

We explored a number of moment method solutions of electromagnetic scattering from simple shapes. I hope the examples presented provided insight into some general scattering mechanisms.

The reader is encouraged to extend the formulations presented and to explore more situations. The solutions for a wire, infinite strips, and a flat plate can be extended to cases where they are bent. The FORTRAN code for scattering from a contour can be modified to analyze scattering from complex shapes. As computational power continues to increase, more and more problems in electromagnetics will succumb to solution with a PC. Computational electromagnetics is still in its dawn—the future is exciting and filled with great promise. It may well produce significant changes to many aspects of how day–to–day electrical engineering is practiced.

# References

[1] Glisson, A.W. and Wilton, D.R., "Simple and Efficient Numerical Methods for Problems of Electromagnetic Radiation and Scattering From Surfaces," *IEEE Transactions on Antennas and Propagation,* Vol. AP–28, No. 5, September 1980, pg. 593–603.

[2] Mahadevan, Auda, and Glisson A.W., "Scattering From a Thin Perfectly-Conducting Square Plate," *IEEE Antennas and Propagation Magazine,* Vol. 34, No. 1, February 1992, pg. 26–32.

[3] Rahmat–Samii, y., and Mittra, Raj, "Integral Equation Solution and RCS Computation of a Thin Rectangular Plate," *IEEE Transactions on Antennas and Propagation,* July 1974, pg. 608–610.

[4] Ross, R.A., "Radar Cross Section of Rectangular Flat Plates as a Function of Aspect Angle," *IEEE Transactions on Antennas and Propagation,* Vol. AP-14, No. 3, May 1966, pg. 329–335.

# Appendixes: FORTRAN Computer Programs

These program listings were compiled using Microsoft FORTRAN. They have been written to comply with FORTRAN 77 standards such that they should easily compile and run using any FORTRAN 77 compiler.* The listings have been typeset directly from the FORTRAN source files using TeX and should contain no typographical errors.

## Chapter 2

CSTRIP.FOR   This program calculates the total charge/unit length of a charged strip of infinite length. The strip is 2m wide. This program implements equations (2.27), (2.28), and (2.29). Page 128.

CRGPLATE.FOR  This program calculates the capacitance of a charged plate. The plate is square and has sides 2m in length. This program implements equation (2.50). Page 135.

## Chapter 3

TRIHALEX.FOR  This program calculates the broadside RCS of a thin wire. The wire's length and radius in wavelengths are provided by the user. The program implements equations (3.41), (3.42), (3.43), (3.44), (3.45), (3.15), (3.6), and (3.7). Page 148.

## Chapter 4

RSTRIPTM.FOR  This program calculates the RCS of a resistive strip for the TM case. The strip is infinitely long and its width is provided by the user. The incident angle of the plane wave and sheet resistance of the strip are also input by the user. The program implements equations (4.8a) and (4.8b). Page 164.

STRIPTE.FOR  This program calculates the RCS of a conductive strip for the TE case. The strip is infinitely long and its width along with the incident

---

* The WATCOM FORTRAN 77 compiler option /cc should be used, for carriage control. This provides screen printout of each matrix element. The programs have also been compiled on a Lahey FORTRAN compiler.

angle of the plane wave are input by the user. The program implements equations (4.21) and (4.22). Page 178.

## Chapter 5

TMCIR.FOR  This program calculates the RCS of a circular contour for the TM case. The contour is infinitely long, and its radius along with the incident angle of the plane wave are input by the user. The program implements equations (5.8) and (5.9). Page 193.

TECIR.FOR  This program calculates the RCS of a circular contour for the TE case. The contour is infinitely long, and its radius along with the incident angle of the plane wave are input by the user. The program implements equations (5.17) and (5.18). Page 208.

## Chapter 6

PLATE.FOR  This program calculates the RCS of a thin metal plate. The plate is square with the size of its sides provided by the user. The angle of the incoming plane wave in $\theta$ and $\phi$ is also input. The program implements equations (6.34) and (6.35). Page 230.

# Appendix A:
# Chapter 2
# FORTRAN Computer Programs

```
C     **********************************************************
C     ** THIS PROGRAM SOLVES FOR THE TOTAL CHARGE PER UNIT  **
C     ** LENGTH OF A LONG STRIP WHICH PRODUCES A 1 VOLT     **
C     ** POTENTIAL.                                         **
C     **                                                    **
C     ** THIS PROGRAM IMPLEMENTS EQUATIONS (2.27),(2.28)    **
C     **                                   AND (2.29)       **
C     **                                                    **
C     **   RANDY BANCROFT  8/19/95                          **
C     **********************************************************
C     **                                                    **
C     ** VARIABLE DICTIONARY (MATRIX SECTION NOT INCLUDED)  **
C     **                                                    **
C     ** A -- MOMENT MATRIX                                 **
C     ** B -- ENFORCEMENT MATRIX (BECOMES SOLUTION MATRIX)  **
C     ** C1 -- CONSTANT USED IN RICHARDSON'S EXTRAPOLATION  **
C     ** CHARGE -- TOTAL CHARGE ON STRIP                    **
C     ** CP -- ARRAY FOR RICHARDSON'S EXTRAPOLATION         **
C     ** EO -- PERMITIVITTY OF FREE SPACE                   **
C     ** G -- GREEN'S FUNCTION                              **
C     ** H -- SEGMENT LENGTH                                **
C     ** I -- LOOP VARIABLE                                 **
C     ** J -- LOOP VARIABLE                                 **
C     ** MTX -- ARRAY SIZE [A],[B]                          **
C     ** NMAX -- MAXIMUM NUMBER OF SEGMENT DIVISIONS        **
C     ** NS -- NUMBER OF SEGMENTS                           **
C     ** PI -- RATIO OF CIRCUMFERENCE TO DIAMETER OF CIRCLE **
C     ** Q -- RICHARDSON'S EXTRAPOLATION LOOP VARIABLE      **
C     ** S -- SEGMENT LOOP COUNTER                          **
C     ** UO -- PERMEABILITY OF FREE SPACE                   **
C     ** XM -- MATCH POINT                                  **
C     ** XN -- LOWER LIMIT OF INTEGRATION OVER SEGMENT N    **
C     ** XN1 -- UPPER LIMIT OF INTEGRATION OVER SEGMENT N   **
C     ** Y -- ARRAY WHICH HOLDS NUMBER OF SEGMENTS FOR S    **
C     **********************************************************
C
      PROGRAM CSTRIP
C
      IMPLICIT NONE
C
      INTEGER MTX
      PARAMETER(MTX=128)
C
C     ***************************
C     ** DECLARE REAL VARIABLES **
C     ***************************
C
      REAL PI,EO,XN,XN1,XM,C1,H,CP(10,10),MAX,G,CHARGE
      REAL A(MTX,MTX),B(MTX),ATEMP,BTEMP,X(MTX),SUM,MJK
      EXTERNAL G
C
C     ***************************
C     ** DECLARE INTEGER VARIBLES **
C     ***************************
```

```
C
      INTEGER K2,ICOL,IROW,COL,XTMP,XINDX(MTX),I,J,N,S,NMAX,Y(10),Q,NS
C
C     ***********************************
C     ** ASSIGN FUNDAMENTAL CONSTANTS **
C     ***********************************
C
      PI=ACOS(-1.0)
      EO=8.854223E-12
C
C     **************************
C     ** MAXIMUM VALUE OF S **
C     ** USED TO DETERMINE   **
C     ** NUMBER OF SEGMENTS **
C     **************************
C
      NMAX=7
C
C     ***********************************
C     ** OUTER LOOP CONTROLS HOW MANY **
C     ** SEGMENT DIVISIONS OCCUR      **
C     ***********************************
C
      DO S=1,NMAX
C
C     **************************
C     ** CALCULATE THE NUMBER **
C     ** OF SEGMENTS FOR THIS **
C     ** LOOP ITERATION       **
C     **************************
C
        NS=2**S
C
C     ****************************
C     ** CALCULATE STEP LENGTH **
C     ****************************
C
        H=2.0/FLOAT(NS)
C
C     *********************************
C     ** FILL THE [A] AND [B] MATRIX **
C     *********************************
C
        DO I=1,NS
          DO J=1,NS
C
C         **************************
C         ** CALCULATE MATCH POINT **
C         **************************
C
          XM= FLOAT(I-1)*H+H/2.0 -1.0
C
C         ***********************************
C         ** CALCULATE XN AND XN+1 WHICH **
C         ** ARE UPPER AND LOWER LIMITS  **
```

```
C          ** OF CURRENT INTEGRATION      **
C          ** OVER THE PRESENT SEGMENT    **
C          ********************************
C
           XN = -1.0 + FLOAT(J-1)*H
           XN1= -1.0 + FLOAT(J)*H
C
C
C          **********************************
C          ** EVALUATE INTEGRATED GREEN'S **
C          ** FUNCTION AT UPPER AND LOWER **
C          ** LIMITS                      **
C          **********************************
C
           A(I,J)=(G(XN1,XM)-G(XN,XM))/(2.0*PI*EO)
C
C          **********************
C          ** PRINT [A] MATRIX **
C          **********************
C
C
C          ********************************
C          ** PRINT THE [A] MATRIX VALUE **
C          ********************************
           WRITE (*,'(A1,4X,A,I3,I3,A,F17.2)')')'+','A(',I,J,')=',A(I,J)
C
           ENDDO
C
C          ***********************
C          ** FILL THE B MATRIX **
C          ***********************
C
           B(I)=1.0
C
         ENDDO
C
C        ********************************************
C        ** SET DUMMY VARIBLES FOR MATRIX ROUTINE **
C        ********************************************
C
         N=NS
C
C        ******************************
C        ** SOLVE FOR CHARGE ON STRIP **
C        ******************************
C
C        **************************************************************
C        ** GAUSSIAN ELIMINATION ALGORITHM WITH TOTAL PIVOTING **
C        ** [B] MATRIX IS REPLACED WITH VALUES OF CHARGE        **
C        **************************************************************
C
C        ************************************
C        ** INITIALIZE SOLUTION INDEX MATRIX **
C        ************************************
C
         DO I=1,N
           XINDX(I)=I
```

```
      END DO
C
C
C     ************************************
C     ** OUTER LOOP CONTROLS ELIMINATION **
C     ************************************
C
      DO K2=1,N-1
C
C     ***************************************
C     ** SEARCH FOR MAXIMUM VALUE IN ARRAY **
C     ***************************************
C
      MAX=0.0
C
      DO I=K2,N
        DO J=K2,N
          IF(ABS(A(I,J)).GT.MAX)THEN
            MAX=A(I,J)
C           ***********************************
C           ** KEEP INDICES OF MAXIMUM ELEMENT **
C           ***********************************
            IROW=I
            ICOL=J
          ENDIF
        END DO
      END DO
C
C     *********************************************************
C     ** DETERMINE IF A ROW EXCHANGE OR COLUMN EXCHANGE **
C     ** IS REQUIRED TO BRING ELEMENT TO PIVOT          **
C     *********************************************************
C
C
C     *********************************************************
C     ** IF THE COLUMN INDEX AND THE ROW INDEX MATCH **
C     ** A ROW EXCHANGE WILL BRING THE MAXIMUM        **
C     ** ELEMENT TO THE PIVOT POINT                   **
C     *********************************************************
      IF(IROW.NE.K2)THEN
C       *********************************************************
C       ** EXCHANGE IROW (I.E. MAX ELEMENT ROW) WITH ROW K2 **
C       *********************************************************
        DO I=1,N
C         ******************************
C         ** FIRST SWAP THE [A] MATRIX **
C         ******************************
          ATEMP=A(IROW,I)
          A(IROW,I)=A(K2,I)
          A(K2,I)=ATEMP
        END DO
C       ******************************
C       ** THEN SWAP THE [B] MATRIX **
C       ******************************
        BTEMP=B(IROW)
        B(IROW)=B(K2)
```

```
          B(K2)=BTEMP
        ENDIF
C
C
C       **************************************************************
C       ** EXCHANGE ICOL (I.E. MAX ELEMENT COLUMN WITH COLUMN K2 **
C       **************************************************************
        IF(ICOL.NE.K2)THEN
C
          DO I=1,N
C               ********************************
C               ** FIRST SWAP THE [A] MATRIX **
C               ********************************
                ATEMP=A(I,ICOL)
                A(I,ICOL)=A(I,K2)
                A(I,K2)=ATEMP
          END DO
C               ***********************************
C               ** THEN SWAP THE [XINDX] MATRIX **
C               ** WHICH DOES THE SOLUTION ORDER **
C               ** BOOK KEEPING                **
C               ***********************************
          XTMP=XINDX(ICOL)
          XINDX(ICOL)=XINDX(K2)
          XINDX(K2)=XTMP
C
        ENDIF
C
C
C       **************************************
C       ** NORMALIZE EACH ROW AND ELIMINATE **
C       **************************************
C
        IF(ABS(A(K2,K2)).EQ.0.0) WRITE(*,*) 'ZERO PIVOT ENCOUNTERED: ERR
     ;OR1'
C
          DO J=K2+1,N
            MJK=A(J,K2)/A(K2,K2)
            DO COL=K2,N
             A(J,COL)=A(J,COL)-MJK*A(K2,COL)
            END DO
            B(J)=B(J)-MJK*B(K2)
          END DO
C
        ENDDO
C
C       ***********************
C       ** BACKSUBSTITUTION **
C       ***********************
C
        X(N)=B(N)/A(N,N)
C
        DO I=(N-1),1,-1
          SUM=0.0
          DO J=(I+1),N
            SUM=SUM+A(I,J)*X(J)
```

```
      END DO
        X(I)=(B(I)-SUM)/A(I,I)
      END DO
C
C
C     **********************************************************
C     ** USE THE INDEX ARRAY TO LOCATE PERMUTATED SOLUTIONS **
C     **********************************************************
C
      DO I=1,N
        DO J=1,N
          IF(XINDX(J).EQ.I) B(I)=X(J)
        END DO
      END DO
C
C
C     ********************************************************
C     ** SUM TO OBTAIN TOTAL CHARGE/UNIT LENGTH **
C     ********************************************************
      CHARGE=0.0
C
      DO I=1,NS
        CHARGE=CHARGE+B(I)
      END DO
C
      CHARGE=CHARGE*H
C
C
C     **********************************************************
C     ** ASSIGN STRIP CAPACITANCE (IN PICOFARADS)          **
C     ** TO AN ARRAY FOR USE IN A RICHARDSON EXTRAPOLATION **
C     **********************************************************
C
      CP(1,S)=CHARGE/1.0E-12
C
C     ********************************************************
C     ** SAVE VALUE OF NUMBER OF SEGMENTS TO [Y] ARRAY **
C     ********************************************************
C
      Y(S)=NS
C
C     ******************************************************
C     ** RICHARDSON EXTRAPOLATION CALCULATION **
C     ******************************************************
C
      C1=1
C
      DO Q=2,S
        C1=2*C1
        CP(Q,S)=(C1*CP(Q-1,S)-CP(Q-1,S-1))/(C1-1.0)
      ENDDO
C
C     **********************************************************
C     ** PRINT OUT CALCULATED TOTAL CHARGE/UNIT LENGTH VALUES **
C     **********************************************************
C
      IF (S.EQ.1) THEN
        DO I=1,25
```

```
            WRITE (*,*) ' '
          ENDDO
        END IF
C
        IF (S.EQ.1) THEN
          WRITE (*,'(12X,A)')'Moment Method Solution for the Total'
          WRITE (*,'(12X,A)')'Charge/Unit Length of a Strip '

          WRITE (*,'(12X,A)')'(Closed Form Pulses With Point Matching)'

          WRITE (*,*) ' '
        END IF
C
        WRITE(*,'(A1,A)')'+',
     ;      '
C
        IF (S.EQ.1) WRITE (*,'(8X,A,9X,A,9X,A)')'M','CHARGE',
     ;      'EXTRAPOLATION'
C
        IF (S.EQ.1) WRITE (*,'(20X,A,16X,A)')'pC','pC'
C
        IF (S.EQ.1) WRITE (*,*) ' '
        IF (S.EQ.1) WRITE (*,*) ' '
C
        WRITE (*,'(A1,5X,I3,3X,F12.4,6X,F12.4)')'+',Y(S),CP(1,S),CP(S,S)
C
        WRITE (*,*) ' '
C
        WRITE (*,*) ' '
C
      ENDDO
C
      END
CC
C     **********************************
C     ** INTEGRATED GREEN'S FUNCTION **
C     **********************************
C
      REAL FUNCTION G(X,XPRM)
      IMPLICIT NONE
      REAL X,XPRM
C
      G=-((X-XPRM)*LOG(ABS(X-XPRM))-(X-XPRM))
C
      RETURN
      END
```

```
C     *************************************************************
C     ** THIS PROGRAM CALCULATES THE CHARGE/UNIT AREA          **
C     ** AND CAPACITANCE OF A SQUARE METAL PLATE WITH          **
C     ** AN APPLIED POTENTIAL OF 1 VOLT                        **
C     ** (SYMMETRY IS NOT USED)                                **
C     **                                                       **
C     ** THIS PROGRAM IMPLEMENTS (2.50)                        **
C     **                                                       **
C     ** RANDY BANCROFT  8/19/95                               **
C     **                                                       **
C     *************************************************************
C     **                                                       **
C     ** VARIABLE DICTIONARY (MATRIX SECTION NOT INCLUDED)     **
C     **                 (INTEGRATION SECTION NOT INCLUDED)    **
C     **                                                       **
C     ** A -- MOMENT MATRIX                                    **
C     ** B -- ENFORCEMENT MATRIX (BECOMES SOLUTION MATRIX)     **
C     ** C1 -- CONSTANT USED IN RICHARDSON'S EXTRAPOLATION     **
C     ** CHARGE -- TOTAL CHARGE ON STRIP                       **
C     ** CP -- ARRAY FOR RICHARDSON'S EXTRAPOLATION            **
C     ** EO -- PERMITIVITTY OF FREE SPACE                      **
C     ** H -- SEGMENT LENGTH                                   **
C     ** I -- LOOP VARIABLE                                    **
C     ** INTX -- INTEGRATION LOOP VARIABLE X COUNTER           **
C     ** INTY -- INTEGRATION LOOP VARIABLE Y COUNTER           **
C     ** J -- LOOP VARIABLE                                    **
C     ** KERNEL -- KERNEL OF INTEGRAL EQUATION                 **
C     **           (AKA GREEN'S FUNCTION)                      **
C     ** MTX -- ARRAY SIZE [A],[B]                             **
C     ** NMAX -- MAXIMUM NUMBER OF SEGMENT DIVISIONS           **
C     ** NS -- NUMBER OF SEGMENTS                              **
C     ** PI -- RATIO OF CIRCUMFERENCE TO DIAMETER OF CIRCLE    **
C     ** Q -- RICHARDSON'S EXTRAPOLATION LOOP VARIABLE         **
C     ** S -- SEGMENT LOOP COUNTER                             **
C     ** XM -- MATCH POINT COORDINATE ALONG X-AXIS             **
C     ** XN -- LOWER LIMIT OF INTEGRATION OVER SEGMENT N       **
C     **         ALONG X-AXIS                                  **
C     ** XN1 -- UPPER LIMIT OF INTEGRATION OVER SEGMENT N      **
C     **         ALONG X-AXIS                                  **
C     ** Y -- ARRAY WHICH HOLDS NUMBER OF SEGMENTS FOR S       **
C     ** YM -- MATCH POINT COORDINATE ALONG Y-AXIS             **
C     ** YN -- LOWER LIMIT OF INTEGRATION OVER SEGMENT N       **
C     **         ALONG Y-AXIS                                  **
C     ** YN1 -- UPPER LIMIT OF INTEGRATION OVER SEGMENT N      **
C     **         ALONG Y-AXIS                                  **
C     *************************************************************
C
      PROGRAM CRGPLATE
C
      IMPLICIT NONE
C
      INTEGER MTX
      PARAMETER(MTX=256)
C
```

```
C      ****************************
C      ** DECLARE REAL VARIABLES **
C      ****************************
C
       REAL PI,EO,XN,XN1,XM,YN,YN1,YM,C1,H,CP(16,16),MAX,MINT2D
       REAL CHARGE,A(MTX,MTX),B(MTX),ATEMP,BTEMP,X(MTX),SUM,MJK
C
       EXTERNAL KERNEL
C
C      *****************************
C      ** DECLARE INTEGER VARIBLES **
C      *****************************
C
       INTEGER K2,ICOL,IROW,COL,XTMP,XINDX(MTX),I,J,N,M,S,NMAX,Y(16),Q,NS
       INTEGER MATCHX,MATCHY,INTX,INTY
C
C      ********************************
C      ** ASSIGN FUNDAMENTAL CONSTANTS **
C      ********************************
C
       PI=ACOS(-1.0)
       EO=8.854223E-12
C
C      *************************
C      ** MAXIMUM VALUE OF S **
C      ** USED TO DETERMINE  **
C      ** NUMBER OF SEGMENTS **
C      *************************
C
       NMAX=4
C
C      *********************************
C      ** OUTER LOOP CONTROLS HOW MANY **
C      ** SEGMENT DIVISIONS OCCUR    **
C      *********************************
C
       DO S=1,NMAX
C
C         ***************************
C         ** CALCULATE THE NUMBER **
C         ** OF SEGMENTS FOR THIS **
C         ** LOOP ITERATION      **
C         ***************************
C
          NS=2**(S)
C
C         ****************************
C         ** CALCULATE STEP LENGTH  **
C         ** ON 2 METER SQUARE PLATE **
C         ****************************
C
          H=2.0/FLOAT(NS)
C
C         *******************************
C         ** FILL THE [A] AND [B] MATRIX **
```

```
C       ********************************
C
C       ********************************
C       ** COUNTER FOR EACH MATCH POINT **
C       ********************************
        M=0
C
        DO MATCHY=1,NS
          DO MATCHX=1,NS
C
C         ***************************************
C         ** CALCULATE X AND Y COORDINATES OF **
C         ** CURRENT MATCH POINT (Xm,Ym)       **
C         ***************************************
C
          XM= -1.0 + FLOAT(MATCHX-1)*H+H/2.0
          YM= -1.0 + FLOAT(MATCHY-1)*H+H/2.0
C
C         *****************************
C         ** NUMBER THE MATCH POINTS **
C         *****************************
          M=M+1
C
C         ***************************************
C         ** CALCULATE THE INTEGRATION LIMITS **
C         ** FOR EACH SEGMENT OF CHARGE        **
C         ***************************************
C
          N=0
C
          DO INTY=1,NS
            DO INTX=1,NS
C
              XN= -1.0 + FLOAT(INTX-1)*H
              XN1= -1.0 + FLOAT(INTX)*H
C
              YN= -1.0 + FLOAT(INTY-1)*H
              YN1= -1.0 + FLOAT(INTY)*H
C
            N=N+1
C
C           **********************************
C           ** EVALUATE INTEGRATED GREEN'S **
C           ** FUNCTION AT UPPER AND LOWER **
C           ** LIMITS                       **
C           **********************************
C
            A(M,N)=MINT2D(KERNEL,XN,XN1,YN,YN1,XM,YM)/(4.0*PI*EO)
C
C           *********************
C           ** PRINT [A] MATRIX **
C           *********************
            WRITE (*,'(A1,4X,A,I3,I3,A,F17.2)')')'+','A(',M,N,')=',A(M,N)
C
            ENDDO
```

```
          ENDDO
C
          ENDDO
       ENDDO
C
C           ***********************
C           ** FILL THE B MATRIX **
C           ***********************
C
       DO I=1,(NS*NS)
         B(I)=1.0
       ENDDO
C
C      *******************************************
C      ** SET DUMMY VARIBLES FOR MATRIX ROUTINE **
C      *******************************************
C
       N=(NS*NS)
       M=(NS*NS)
C
C      *****************************************************
C      ** SOLVE FOR PLATE CHARGE USING GAUSS FULL PIVOT **
C      *****************************************************
C
C      **********************************************************
C      ** GAUSSIAN ELIMINATION ALGORITHM WITH TOTAL PIVOTING **
C      **********************************************************
C
C      **************************************
C      ** INITIALIZE SOLUTION INDEX MATRIX **
C      **************************************
C
       DO I=1,N
         XINDX(I)=I
       END DO
C
C
C      ***********************************
C      ** OUTER LOOP CONTROLS ELIMINATION **
C      ***********************************
C
       DO K2=1,N-1
C
C      **************************************
C      ** SEARCH FOR MAXIMUM VALUE IN ARRAY **
C      **************************************
C
       MAX=0.0
C
       DO I=K2,N
         DO J=K2,N
           IF(ABS(A(I,J)).GT.MAX)THEN
             MAX=A(I,J)
C            ***********************************
C            ** KEEP INDICES OF MAXIMUM ELEMENT **
```

```
C          **************************************
           IROW=I
           ICOL=J
         ENDIF
       END DO
     END DO
C
C      ********************************************************
C      ** DETERMINE IF A ROW EXCHANGE OR COLUMN EXCHANGE **
C      ** IS REQUIRED TO BRING ELEMENT TO PIVOT          **
C      ********************************************************
C
C      ********************************************************
C      ** IF THE COLUMN INDEX AND THE ROW INDEX MATCH **
C      ** A ROW EXCHANGE WILL BRING THE MAXIMUM        **
C      ** ELEMENT TO THE PIVOT POINT                   **
C      ********************************************************
       IF(IROW.NE.K2)THEN
C         ****************************************************
C         ** EXCHANGE IROW (I.E. MAX ELEMENT ROW) WITH ROW K2 **
C         ****************************************************
          DO I=1,N
C            *****************************
C            ** FIRST SWAP THE [A] MATRIX **
C            *****************************
             ATEMP=A(IROW,I)
             A(IROW,I)=A(K2,I)
             A(K2,I)=ATEMP
          END DO
C            *****************************
C            ** THEN SWAP THE [B] MATRIX **
C            *****************************
             BTEMP=B(IROW)
             B(IROW)=B(K2)
             B(K2)=BTEMP
       ENDIF
C
C      ****************************************************************
C      ** EXCHANGE ICOL (I.E. MAX ELEMENT COLUMN WITH COLUMN K2 **
C      ****************************************************************
       IF(ICOL.NE.K2)THEN
C
          DO I=1,N
C            *****************************
C            ** FIRST SWAP THE [A] MATRIX **
C            *****************************
             ATEMP=A(I,ICOL)
             A(I,ICOL)=A(I,K2)
             A(I,K2)=ATEMP
          END DO
C            *********************************
C            ** THEN SWAP THE [XINDX] MATRIX **
C            ** WHICH DOES THE SOLUTION ORDER **
C            ** BOOK KEEPING                  **
C            *********************************
```

```
          XTMP=XINDX(ICOL)
          XINDX(ICOL)=XINDX(K2)
          XINDX(K2)=XTMP
C
      ENDIF
C
C
C     **************************************
C     ** NORMALIZE EACH ROW AND ELIMINATE **
C     **************************************
C
      IF(ABS(A(K2,K2)).EQ.0.0) WRITE(*,*) 'ZERO PIVOT ENCOUNTERED: ERROR
     ;1'
C
      DO J=K2+1,N
        MJK=A(J,K2)/A(K2,K2)
        DO COL=K2,N
        A(J,COL)=A(J,COL)-MJK*A(K2,COL)
        END DO
        B(J)=B(J)-MJK*B(K2)
      END DO
C
      ENDDO
C
C     ***********************
C     ** BACKSUBSTITUTION **
C     ***********************
C
      X(N)=B(N)/A(N,N)
C
      DO I=(N-1),1,-1
        SUM=0.0
        DO J=(I+1),N
          SUM=SUM+A(I,J)*X(J)
        END DO
        X(I)=(B(I)-SUM)/A(I,I)
      END DO
C
C     *************************************************************
C     ** USE THE INDEX ARRAY TO LOCATE PERMUTATED SOLUTIONS **
C     *************************************************************
C
      DO I=1,N
        DO J=1,N
          IF(XINDX(J).EQ.I) B(I)=X(J)
        END DO
      END DO
C
C     ********************************************
C     ** INTEGRATE CHARGE TO OBTAIN CAPACITANCE **
C     ********************************************
C
      CHARGE=0.0
C
      DO I=1,(NS*NS)
```

```
        CHARGE=CHARGE+B(I)
      END DO
C
      CHARGE=CHARGE*H*H

C
C       *****************************************
C       ** ASSIGN CAPACITANCE TO AN ARRAY FOR **
C       ** USE IN A RICHARDSON EXTRAPOLATION  **
C       **(OUTPUT IN PICOFARADS)              **
C       *****************************************
C
      CP(1,S)=CHARGE/1.0E-12

C
C       *****************************************
C       ** SAVE NUMBER OF SEGMENTS TO [Y] ARRAY **
C       *****************************************
C
      Y(S)=NS

C
C       *****************************************
C       ** RICHARDSON EXTRAPOLATION CALCULATION **
C       *****************************************
C
      C1=1
C
      DO Q=2,S
        C1=2*C1
        CP(Q,S)=(C1*CP(Q-1,S)-CP(Q-1,S-1))/(C1-1.0)
      ENDDO
C
C       *****************************************
C       ** PRINT OUT CALCULATED VALUES OF **
C       ** PLATE CAPACITANCE              **
C       *****************************************
C
      IF (S.EQ.1) THEN
        DO I=1,25
          WRITE (*,*) ' '
        ENDDO
      END IF
C
      IF (S.EQ.1) THEN
        WRITE (*,'(12X,A)')'Moment Method Solution for the'
        WRITE (*,'(12X,A)')'Capacitance of Charged 2 meter Plate'

        WRITE (*,'(12X,A)')'(Pulses With Point Matching No Symmetry)'

        WRITE (*,'(12X,A)')'(Numerical Integration)'

        WRITE (*,*) ' '
        WRITE (*,*) ' '
      END IF
C
      WRITE(*,'(A1,A)')'+',' '
```

```
      ;                '
C
          IF (S.EQ.1) WRITE (*,'(8X,A,11X,A,11X,A)')'M','CAP',
      ;         'EXTRAPOLATION'
C
          IF (S.EQ.1) WRITE (*,'(16X,A,8X,A)')'PICOFARADS','PICOFARADS'
C
          IF (S.EQ.1) WRITE (*,*) ' '
C
          WRITE (*,'(A1,5X,I3,3X,F12.4,6X,F12.4)')'+',Y(S),CP(1,S),CP(S,S)

          WRITE (*,*) ' '
          WRITE (*,*) ' '
C
      ENDDO
C
      END
CC
C     *********************************
C     ** KERNEL OF INTEGRAL EQUATION **
C     *********************************
C
      REAL FUNCTION KERNEL(X,XPRM,Y,YPRM)
      IMPLICIT NONE
      REAL X,XPRM,Y,YPRM
C
      KERNEL=1.0/SQRT((X-XPRM)*(X-XPRM)+(Y-YPRM)*(Y-YPRM))
C
      RETURN
      END
CC
C
C     **************************************************************
C     ** THIS SUBROUTINE APPROXIMATES THE DOUBLE INTEGRAL OF A **
C     ** FUNCTION USING MIDPOINT INTEGRATION WITH RICHARDSON   **
C     ** EXTRAPOLATION                                          **
C     **************************************************************
C
      REAL FUNCTION MINT2D(F,A,B,C,D,XM,YM)
C
      IMPLICIT NONE
      REAL CP(16,16),C1,SUMX,F
      REAL NSX,DX,A,B,MIDPNT,ERROR,OLD,NEW,TOL,C,D,MIDINT,XM,YM
      INTEGER COUNTX,I,Q,NMAX
      LOGICAL FINISHED
      EXTERNAL F
C
      FINISHED=.FALSE.
      COUNTX=1
      TOL=5.0E-5
      NMAX=16
C
      DO WHILE(.NOT.FINISHED)
C
C         ***********************************
```

```
C          ** CALCULATE NUMBER OF SUBSECTIONS **
C          **********************************
           NSX=2.0**COUNTX
C
C          *************************
C          ** CALCULATE STEP-SIZE **
C          *************************
           DX=(B-A)/NSX
C
C          *********************************************
C          ** EVALUATE AND SUM FUNCTION VALUE AT MIDPOINTS **
C          ** FUNCTION IS FIRST INTEGRATION              **
C          *********************************************
C
           SUMX=0.0
C
           DO I=1,NSX
             MIDPNT=A+FLOAT(I-1)*DX+DX/2.0
             SUMX=SUMX+MIDINT(F,C,D,MIDPNT,XM,YM)
           ENDDO
C
C
C          *********************************************
C          ** ASSIGN SUMATION TO ARRAY FOR EXTRAPOLATION **
C          *********************************************
           CP(1,COUNTX)=SUMX*DX
C
C          *******************************************
C          ** RICHARDSON EXTRAPOLATION CALCULATION **
C          *******************************************
C
           C1=1.0
C
           DO Q=2,COUNTX
             C1=2*C1
             CP(Q,COUNTX)=(C1*CP(Q-1,COUNTX)-CP(Q-1,COUNTX-1))/(C1-1.0)
           ENDDO
C
           MINT2D=CP(COUNTX,COUNTX)
C
           NEW=CP(COUNTX,COUNTX)
C
C          ***********************************
C          ** CALCULATE ERROR AFTER FIRST PASS **
C          ***********************************
C
           IF((COUNTX.GT.1).AND.(ABS(OLD-NEW).NE.0.0)) THEN
             ERROR=ABS(OLD-NEW)/ABS(NEW)
           ENDIF
C
           IF(ABS(OLD-NEW).EQ.0.0) FINISHED=.TRUE.
C
C          *************************************************
C          ** CHECK ERROR AFTER FIRST PASS TO SEE IF INTEGRATION **
C          ** IS WITHIN TOLERANCE END CALCULATION IF IT IS       **
```

```
C
C       ************************************************************
C
        IF((COUNTX.GT.1).AND.(ERROR.LT.TOL)) FINISHED=.TRUE.
C
C       ************************************************************
C       ** CHECK TO SEE IF WE HAVE REACHED THE MAXIMUM NUMBER **
C       ** OF SUBSECTIONS ALLOWED BY USER                     **
C       ************************************************************
C
        IF(COUNTX.EQ.NMAX) FINISHED=.TRUE.
C
        COUNTX=COUNTX+1
C
        OLD=NEW
C
C
      END DO
C
      RETURN
      END
C
C
C       ****************************************************************
C       ** THIS SUBROUTINE APPROXIMATES THE INTEGRAL OF A FUNCTION  **
C       ** USING MIDPOINT INTEGRATION WITH RICHARDSON EXTRAPOLATION **
C       ****************************************************************
C
        REAL FUNCTION MIDINT(F,C,D,X,XM,YM)
C
        IMPLICIT NONE
        REAL CP(16,16),C1,SUMX,F
        REAL NSX,DX,C,D,XM,YM,X,MIDPNT,ERROR,OLD,NEW,TOL
        INTEGER COUNTX,I,Q,NMAX
        LOGICAL FINISHED
        EXTERNAL F
C
        FINISHED=.FALSE.
        COUNTX=1
        TOL=5.0E-5
        NMAX=16
C
        DO WHILE(.NOT.FINISHED)
C
C         ***************************************
C         ** CALCULATE NUMBER OF SUBSECTIONS **
C         ***************************************
          NSX=2.0**COUNTX
C
C         *************************
C         ** CALCULATE STEP-SIZE **
C         *************************
          DX=(D-C)/NSX
C
C         ****************************************************
```

```
C          ** EVALUATE AND SUM FUNCTION VALUE MID/POINTS **
C          *************************************************
C
           SUMX=0.0
C
           DO I=1,NSX
             MIDPNT=C+FLOAT(I-1)*DX+DX/2.0
               SUMX=SUMX+F(XM,X,YM,MIDPNT)
           ENDDO
C
C
C          **************************************************
C          ** ASSIGN SUMATION TO ARRAY FOR EXTRAPOLATION **
C          **************************************************
           CP(1,COUNTX)=SUMX*DX
C
C          ******************************************
C          ** RICHARDSON EXTRAPOLATION CALCULATION **
C          ******************************************
C
           C1=1.0
C
           DO Q=2,COUNTX
             C1=2*C1
             CP(Q,COUNTX)=(C1*CP(Q-1,COUNTX)-CP(Q-1,COUNTX-1))/(C1-1.0)
           ENDDO
C
           MIDINT=CP(COUNTX,COUNTX)
C
           NEW=CP(COUNTX,COUNTX)
C
C          ************************************
C          ** CALCULATE ERROR AFTER FIRST PASS **
C          ************************************
C
           IF(COUNTX.GT.1) ERROR=ABS(OLD-NEW)/ABS(NEW)
C
C          *********************************************************
C          ** CHECK ERROR AFTER FIRST PASS TO SEE IF INTEGRATION  **
C          ** IS WITHIN TOLERANCE END CALCULATION IF IT IS        **
C          *********************************************************
C
           IF((COUNTX.GT.1).AND.(ERROR.LT.TOL)) FINISHED=.TRUE.
C
C          *********************************************************
C          ** CHECK TO SEE IF WE HAVE REACHED THE MAXIMUM NUMBER  **
C          ** OF SUBSECTIONS ALLOWED BY USER                      **
C          *********************************************************
C
           IF(COUNTX.EQ.NMAX) FINISHED=.TRUE.
C
           COUNTX=COUNTX+1
C
           OLD=NEW
C
```

```
C
      END DO
C
      RETURN
      END
```

# Appendix B:
## Chapter 3
## FORTRAN Computer Programs

```
C    *********************************************************
C    ** THIS PROGRAM USES TRIANGULAR EXPANSION              **
C    ** FUNCTIONS TO SOLVE HALLEN'S                         **
C    ** EQUATION USING THE METHOD OF MOMENTS.               **
C    ** THE MATCH POINT IS AT THE CENTER OF                 **
C    ** EACH TRIANGLE. THE FIELD FROM THE                   **
C    ** CURRENT IS USED TO CALCULATE THE RADAR              **
C    ** CROSS SECTION (RCS) THIS PROGRAM USES               **
C    ** SYMMETRY AND THE EXACT KERNEL                       **
C    **                                                     **
C    ** THIS PROGRAM IMPLEMENTS (3.41,3.42,3.43,3.44,3.45)  **
C    **                                                     **
C    ** RANDY BANCROFT 8/19/95                              **
C    **                                                     **
C    *********************************************************
C    **                                                     **
C    ** VARIABLE DICTIONARY (MATRIX SECTION NOT INCLUDED)   **
C    **               (INTEGRATION SECTION NOT INCLUDED)    **
C    **                                                     **
C    ** A -- MOMENT MATRIX                                  **
C    ** B -- ENFORCEMENT MATRIX (BECOMES SOLUTION MATRIX)   **
C    ** C -- SPEED OF LIGHT IN FREE SPACE                   **
C    ** C1 -- CONSTANT USED IN RICHARDSON'S EXTRAPOLATION   **
C    ** CP -- ARRAY FOR RICHARDSON'S EXTRAPOLATION          **
C    ** CSUM -- COMPLEX VARIBLE FOR A SUMMATION             **
C    ** EO -- PERMITIVITTY OF FREE SPACE                    **
C    ** ELEMENT -- TEMPORARY VARIABLE FOR A(I,J)            **
C    ** ETA -- CHARACTERISTIC IMPEDANCE OF FREE SPACE       **
C    ** FREQ -- FREQUENCY OF 1 METER WAVE                   **
C    ** G -- GREEN'S FUNCTION                               **
C    ** H -- SEGMENT LENGTH                                 **
C    ** I -- LOOP VARIABLE                                  **
C    ** J -- LOOP VARIABLE                                  **
C    ** K -- FREE SPACE WAVENUMBER                          **
C    ** JAY -- SQUARE ROOT OF -1                            **
C    ** L -- LENGTH OF WIRE                                 **
C    ** LAMBDA -- FREE SPACE WAVELENGTH                     **
C    ** LWV -- LENGTH OF WIRE IN WAVELENGTHS                **
C    ** MTX -- ARRAY SIZE [A],[B]                           **
C    ** NMAX -- MAXIMUM NUMBER OF SEGMENT DIVISIONS         **
C    ** NS -- NUMBER OF SEGMENTS                            **
C    ** OMEGA -- ANGULAR FREQUENCY OF WAVE                  **
C    ** PI -- RATIO OF CIRCUMFERENCE TO DIAMETER OF CIRCLE  **
C    ** Q -- RICHARDSON'S EXTRAPOLATION LOOP VARIABLE       **
C    ** R -- RADIUS OF WIRE IN WAVELENGTHS                  **
C    ** S -- SEGMENT LOOP COUNTER                           **
C    ** SIGMA -- NORMALIZED RADAR CROSS SECTION             **
C    ** UO -- PERMEABILITY OF FREE SPACE                    **
C    ** ZM -- MATCH POINT                                   **
C    ** ZN -- LOWER LIMIT OF INTEGRATION OVER SEGMENT N     **
C    ** ZN1 -- UPPER LIMIT OF INTEGRATION OVER SEGMENT N    **
C    ** Y -- ARRAY WHICH STORES SEGMENT NUMBER FOR EACH S   **
C    *********************************************************
C
```

```
C
C
      PROGRAM TRIHALEX
C
      IMPLICIT NONE
C
      INTEGER MTX
      PARAMETER(MTX=129)
C
C     ****************************
C     ** DECLARE REAL VARIABLES **
C     ****************************
C
      REAL EO,UO,C,PI,ZN,ZN1,ZM,LAMBDA,K,L,C1,H
      REAL FREQ,OMEGA,SIGMA,CP(10,10),RADIUS,LWV,MAX
C
C     ****************************
C     ** DECLARE COMPLEX VARIBLES **
C     ****************************
C
      COMPLEX JAY,A(MTX,MTX),B(MTX),ELEMENT,CSUM
      COMPLEX ATEMP,BTEMP,X(MTX),SUM,MJK
C
C     ****************************
C     ** DECLARE INTEGER VARIBLES **
C     ****************************
C
      INTEGER I,J,N,S,NMAX,Y(10),Q,NS
      INTEGER K2,ICOL,IROW,COL,XTMP,XINDX(MTX)
C
C     ****************************
C     ** INPUT DESIRED OPTIONS **
C     ****************************
C
      WRITE (*,'(5X,A)')'INPUT LENGTH OF DIPOLE IN WAVELENGTHS '
      READ (*,*) LWV
      LWV=LWV/2.0
C
C
      WRITE (*,'(5X,A)')'INPUT RADIUS OF DIPOLE IN WAVELENGTHS '
      READ (*,*) RADIUS
C
C     ********************************
C     ** ASSIGN FUNDAMENTAL CONSTANTS **
C     ********************************
C
      JAY=(0.0,1.0)
      C=2.997925E8
      EO=8.854223E-12
      UO=1.256640E-6
      PI=ACOS(-1.0)
C
C     ****************************
C     ** USE 1 METER LAMBDA VALUES **
```

```
C       ******************************
C
        FREQ=C
        OMEGA=2.0*PI*FREQ
        LAMBDA=C/FREQ
        L=LWV*LAMBDA
        K=2.0*PI/LAMBDA
C
C       *************************
C       ** MAXIMUM VALUE OF S **
C       ** USED TO DETERMINE  **
C       ** NUMBER OF SEGMENTS **
C       *************************
C
        NMAX=7
C
C       ********************************
C       ** OUTER LOOP CONTROLS HOW MANY **
C       ** SEGMENT DIVISIONS OCCUR      **
C       ********************************
C
        DO S=1,NMAX
C
C       *************************
C       ** CALCULATE THE NUMBER **
C       ** OF SEGMENTS FOR THIS **
C       ** LOOP ITERATION       **
C       *************************
C
          NS=2**S
C
C       ***************************
C       ** CALCULATE STEP LENGTH **
C       ***************************
C
          H=L/FLOAT(NS)
C
C       *******************************
C       ** FILL THE [A] AND [B] MATRIX **
C       *******************************
C
          DO I=1,NS+1
            DO J=1,NS+1
C
C         ***************************
C         ** CALCULATE MATCH POINT **
C         ***************************
C
              ZM=FLOAT(I-1)*H
C
C         *******************************
C         ** CALCULATE ZN AND ZN+1 WHICH **
C         ** ARE UPPER AND LOWER LIMITS  **
C         ** OF CURRENT INTEGRATION      **
C         ** OVER THE PRESENT SEGMENT    **
```

```
C     ********************************
C
C
C
          IF(J.EQ.1)THEN
            ZN=0.0
            ZN1=H
          ELSE
            ZN=FLOAT(J-2)*H
            ZN1=ZN+2.0*H
          ENDIF

C     ************************************
C     ** IF THE [A] MATRIX ELEMENT IS AT  **
C     ** NS THEN AUGMENT WITH COSINE      **
C     ************************************
C
C
C        ************************************
C        ** EVALUATE INTEGRATED GREEN'S **
C        ** FUNCTION AT UPPER AND LOWER **
C        ** LIMITS                      **
C        ************************************
C
         IF (J.LT.NS+1)THEN

           CALL INTEGRAL (ZN,ZN1,ZM,RADIUS,H,ELEMENT)
           A(I,J)=ELEMENT/(JAY*4.0*PI*EO*OMEGA)*K**2

         ELSE
           A(I,J)=-COS(K*ZM)
         END IF


         **********************
         ** PRINT [A] MATRIX **
         **********************
         WRITE(*,'(A1,4X,A,I3,I3,A,F15.4,2X,F15.4)')
    ;              '+','A(',I,J,')=',A(I,J)

      ENDDO

         **********************
         ** FILL THE B MATRIX **
         **********************

      B(I)=(-1.0,0.0)

   ENDDO

   *********************************************
   ** SET DUMMY VARIBLES FOR MATRIX ROUTINE **
   *********************************************

   N=NS+1

   *********************************************
```

```
C     ** SOLVE FOR CURRENTS USING GAUSS FULL PIVOT **
C     **********************************************

C     *************************************************************
C     ** GAUSSIAN ELIMINATION ALGORITHM WITH TOTAL PIVOTING **
C     *************************************************************
C
C     *************************************
C     ** INITIALIZE SOLUTION INDEX MATRIX **
C     *************************************
C
      DO I=1,N
        XINDX(I)=I
      END DO
C
C
C     ************************************
C     ** OUTER LOOP CONTROLS ELIMINATION **
C     ************************************
C
      DO K2=1,N-1
C
C     ************************************
C     ** SEARCH FOR MAXIMUM VALUE IN ARRAY **
C     ************************************
C
      MAX=0.0
C
      DO I=K2,N
        DO J=K2,N
          IF(CABS(A(I,J)).GT.MAX)THEN
            MAX=A(I,J)
C     ************************************
C     ** KEEP INDICES OF MAXIMUM ELEMENT **
C     ************************************
            IROW=I
            ICOL=J
          ENDIF
        END DO
      END DO
C
C     ****************************************************
C     ** DETERMINE IF A ROW EXCHANGE OR COLUMN EXCHANGE **
C     ** IS REQUIRED TO BRING ELEMENT TO PIVOT          **
C     ****************************************************
C
C     ****************************************************
C     ** IF THE COLUMN INDEX AND THE ROW INDEX MATCH **
C     ** A ROW EXCHANGE WILL BRING THE MAXIMUM        **
C     ** ELEMENT TO THE PIVOT POINT                   **
C     ****************************************************
      IF(IROW.NE.K2)THEN
C     ****************************************************
C     ** EXCHANGE IROW (I.E. MAX ELEMENT ROW) WITH ROW K2 **
C     ****************************************************
```

```
      DO I=1,N
C        ********************************
C        ** FIRST SWAP THE [A] MATRIX **
C        ********************************
         ATEMP=A(IROW,I)
         A(IROW,I)=A(K2,I)
         A(K2,I)=ATEMP
      END DO
C        ********************************
C        ** THEN SWAP THE [B] MATRIX **
C        ********************************
         BTEMP=B(IROW)
         B(IROW)=B(K2)
         B(K2)=BTEMP
      ENDIF
C
C
C     ***********************************************************
C     ** EXCHANGE ICOL (I.E. MAX ELEMENT COLUMN WITH COLUMN K2 **
C     ***********************************************************
      IF(ICOL.NE.K2)THEN
C
         DO I=1,N
C        ********************************
C        ** FIRST SWAP THE [A] MATRIX **
C        ********************************
         ATEMP=A(I,ICOL)
         A(I,ICOL)=A(I,K2)
         A(I,K2)=ATEMP
      END DO
C        *********************************
C        ** THEN SWAP THE [XINDX] MATRIX **
C        ** WHICH DOES THE SOLUTION ORDER **
C        ** BOOK KEEPING                  **
C        *********************************
         XTMP=XINDX(ICOL)
         XINDX(ICOL)=XINDX(K2)
         XINDX(K2)=XTMP
C
      ENDIF
C
C
C
C     ************************************
C     ** NORMALIZE EACH ROW AND ELIMINATE **
C     ************************************
C
      IF(CABS(A(K2,K2)).EQ.0.0) WRITE(*,*) 'ZERO PIVOT ENCOUNTERED: ERRO
     ;R1'
C
      DO J=K2+1,N
        MJK=A(J,K2)/A(K2,K2)
        DO COL=K2,N
         A(J,COL)=A(J,COL)-MJK*A(K2,COL)
        END DO
         B(J)=B(J)-MJK*B(K2)
      END DO
```

```
C
        ENDDO
C
C       ***********************
C       ** BACKSUBSTITUTION **
C       ***********************
C
        X(N)=B(N)/A(N,N)
C
        DO I=(N-1),1,-1
          SUM=(0.0,0.0)
          DO J=(I+1),N
            SUM=SUM+A(I,J)*X(J)
          END DO
          X(I)=(B(I)-SUM)/A(I,I)
        END DO
C
C       **********************************************************
C       ** USE THE INDEX ARRAY TO LOCATE PERMUTATED SOLUTIONS **
C       **********************************************************
C
        DO I=1,N
          DO J=1,N
            IF(XINDX(J).EQ.I) B(I)=X(J)
          END DO
        END DO
C
C       *******************************************
C       ** CACULATE SUM WHICH IS PROPORTIONAL   **
C       ** TO THE SCATTERED FIELD AT 90 DEGREE  **
C       ** INCIDENCE                            **
C       *******************************************
C
        CSUM=(0.0,0.0)
        DO I=1,NS
C         *******************************************
C         ** FACTOR OF TWO COMES FROM SYMMETRY **
C         ** WITH ONLY A HALFPULSE AT THE      **
C         ** CENTER INTEGRATE ONLY HALF        **
C         *******************************************
          IF(I.EQ.1)THEN
            CSUM=CSUM+B(I)*H*JAY*OMEGA*UO
          ELSE
            CSUM=CSUM+2.0*B(I)*H*JAY*OMEGA*UO
          ENDIF
        ENDDO
C
C       *****************************************
C       ** CALCULATE THE RADAR CROSS SECTION **
C       *****************************************
C
        SIGMA=REAL(CSUM*CONJG(CSUM))/(4.0*PI)
C
C       *****************************************
C       ** ASSIGN CURRENT OR TO AN ARRAY FOR  **
```

```
C     ** USE IN A RICHARDSON EXTRAPOLATION   **
C     *********************************************
C
      CP(1,S)=SIGMA
C
C     *********************************************
C     ** SAVE NUMBER OF SEGMENTS  TO [Y] ARRAY **
C     *********************************************
C
      Y(S)=NS
C
C     *********************************************
C     ** RICHARDSON EXTRAPOLATION CALCULATION **
C     *********************************************
C
      C1=1
      DO Q=2,S
        C1=2*C1
        CP(Q,S)=(C1*CP(Q-1,S)-CP(Q-1,S-1))/(C1-1)
      ENDDO
C
      IF (S.EQ.1) THEN
        DO I=1,25
          WRITE (*,*) ' '
        ENDDO
      END IF
C
      IF (S.EQ.1) THEN
        WRITE (*,'(12X,A)')'Hallen Eq.  Triangles with Point Matching (
     ;Exact Kernel):  '
      END IF
C
      IF (S.EQ.1) THEN
        WRITE (*,'(12X,A,F6.3)')'LENGTH OF DIPOLE IN WAVELENGTHS: ',
     ;2.0*LWV
        WRITE (*,'(12X,A,F10.6)')'RADIUS OF DIPOLE IN WAVELENGTHS: ',
     ;RADIUS
        WRITE (*,*) ' '
      END IF
C
C
      IF (S.EQ.1) WRITE(*,'(A1,A)')'+',
     ;         '
C
      IF (S.EQ.1)WRITE (*,'(8X,A,9X,A,7X,A)')'M','(RCS)','EXTRAPOLATIO
     ;N'
C
      IF (S.EQ.1) WRITE (*,'(16X,A,6X,A)')'Sigma/Lambda',
     ;'Sigma/Lambda'
C
      IF (S.EQ.1) WRITE (*,*) ' '
C
      IF (S.EQ.1) WRITE (*,'(A1,A)')'+',
```

```
;             '
      WRITE (*,'(A1,A)')'+',' 
;             '
C
      WRITE (*,'(5X,I3,3X,F12.4,6X,F12.4,15X)')Y(S),CP(1,S),CP(S,S)
C
      WRITE(*,*)' '
C
      ENDDO
C
      END
CC
C     **************************************************
C     ** CALCULATE INTEGRAL OF GREEN'S FUNCTION      **
C     ** USING MIDPOINT INTEGRATION. THE INTEGRATION **
C     ** IS FROM A TO B WITH ZM AS THE MATCH POINT.  **
C     ** INT IS THE RETURNED INTEGRATION VALUE       **
C     **************************************************
C
      SUBROUTINE INTEGRAL (A,B,ZM,AD,H,INT)
C
      IMPLICIT NONE
C
      REAL A,B,ZM,GR,GI,RLINT,IMINT,MIDINT,AD,H
C
      COMPLEX INT
C
      EXTERNAL GR,GI
C
      RLINT=MIDINT(GR,A,B,ZM,AD,H)
      IMINT=MIDINT(GI,A,B,ZM,AD,H)

      INT=CMPLX(RLINT,IMINT)
C
      RETURN
      END
CC
C     ***********************************
C     ** REAL PART OF GREEN'S FUNCTION **
C     ***********************************
C
      REAL FUNCTION GR(Z,ZPRM,AR,H,MID)
      IMPLICIT NONE
      REAL PI,K,Z,ZPRM,R,AR,RPRM,H,MID,TRI
C
      REAL CP(16,16),C1,SUMX
      REAL NSX,DX,A,B,MIDPNT,ERROR,OLD,NEW,TOL
      INTEGER COUNTX,I,Q,NMAX
      LOGICAL FINISHED
C
      PI=ACOS(-1.0)
      K=2.0*PI
      FINISHED=.FALSE.
      COUNTX=1
      TOL=5.0E-5
```

```
      NMAX=16
      A=0.0
      B=2.0*PI

C     DO WHILE(.NOT.FINISHED)

C     *******************************************
C     ** CALCULATE NUMBER OF SUBSECTIONS **
C     *******************************************
C     NSX=2.0**COUNTX
C
C     ****************************
C     ** CALCULATE STEP-SIZE **
C     ****************************
C     DX=(B-A)/NSX
C
C     *******************************************************
C     ** EVALUATE AND SUM FUNCTION VALUE AT MIDPOINTS **
C     *******************************************************
C
      SUMX=0.0
C
      DO I=1,NSX
        MIDPNT=A+FLOAT(I-1)*DX+DX/2.0
        R=SQRT((ZPRM-Z)*(ZPRM-Z)-4.0*(AR*AR)*(SIN(MIDPNT/2)**2))
        RPRM=SQRT((ZPRM+Z)*(ZPRM-Z)+4.0*(AR*AR)*(SIN(MIDPNT/2)**2))
        SUMX=SUMX+(COS(K*R)/R+COS(K*RPRM)/RPRM)
      ENDDO
C
C
C     ***************************************************
C     ** ASSIGN SUMATION TO ARRAY FOR EXTRAPOLATION **
C     ***************************************************
C     CP(1,COUNTX)=SUMX*DX
C
C     *********************************************
C     ** RICHARDSON EXTRAPOLATION CALCULATION **
C     *********************************************
C
      C1=1.0
C
      DO Q=2,COUNTX
        C1=2*C1
        CP(Q,COUNTX)=(C1*CP(Q-1,COUNTX)-CP(Q-1,COUNTX-1))/(C1-1.0)
      ENDDO
C
      GR=CP(COUNTX,COUNTX)
C
      NEW=CP(COUNTX,COUNTX)
C
C     ***************************************
C     ** CALCULATE ERROR AFTER FIRST PASS **
C     ***************************************
C
      IF(COUNTX.GT.1) ERROR=ABS(OLD-NEW)/ABS(NEW)
```

```
C
C
C     ***********************************************************
C     ** CHECK ERROR AFTER FIRST PASS TO SEE IF INTEGRATION  **
C     ** IS WITHIN TOLERANCE END CALCULATION IF IT IS        **
C     ***********************************************************
C
      IF((COUNTX.GT.1).AND.(ERROR.LT.TOL)) FINISHED=.TRUE.
C
C     ***********************************************************
C     ** CHECK TO SEE IF WE HAVE REACHED THE MAXIMUM NUMBER **
C     ** OF SUBSECTIONS ALLOWED BY USER                     **
C     ***********************************************************
C
      IF(COUNTX.EQ.NMAX) FINISHED=.TRUE.
C
      COUNTX=COUNTX+1
C
      OLD=NEW
C
      END DO
C
      GR=GR/(2.0*PI)*TRI(ZPRM,H,MID)
C
      RETURN
      END
CC
CC
C     *****************************************
C     ** IMAGINARY PART OF GREEN'S FUNCTION **
C     *****************************************
C
      REAL FUNCTION GI(Z,ZPRM,AR,H,MID)
      IMPLICIT NONE
      REAL AR,PI,K,Z,ZPRM,R,RPRM,H,MID,TRI
C
      REAL CP(16,16),C1,SUMX
      REAL NSX,DX,A,B,MIDPNT,ERROR,OLD,NEW,TOL
      INTEGER COUNTX,I,Q,NMAX
      LOGICAL FINISHED
C
      PI=ACOS(-1.0)
      K=2.0*PI
      FINISHED=.FALSE.
      COUNTX=1
      TOL=5.0E-5
      NMAX=16
      A=0.0
      B=2.0*PI
C
      DO WHILE(.NOT.FINISHED)
C
C        ***********************************
C        ** CALCULATE NUMBER OF SUBSECTIONS **
C        ***********************************
         NSX=2.0**COUNTX
```

```
C
C
C       *************************
C       ** CALCULATE STEP-SIZE **
C       *************************
        DX=(B-A)/NSX
C
C       ***************************************************
C       ** EVALUATE AND SUM FUNCTION VALUE AT MIDPOINTS **
C       ***************************************************
C
        SUMX=0.0
C
        DO I=1,NSX
          MIDPNT=A+FLOAT(I-1)*DX+DX/2.0
          R=SQRT((ZPRM-Z)*(ZPRM-Z)+4.0*(AR*AR)*(SIN(MIDPNT/2)**2))
          RPRM=SQRT((ZPRM-Z)*(ZPRM+Z)+4.0*(AR*AR)*(SIN(MIDPNT/2)**2))
          SUMX=SUMX-(SIN(K*R)/R+SIN(K*RPRM)/RPRM)
        ENDDO
C
C
C       ***********************************************
C       ** ASSIGN SUMATION TO ARRAY FOR EXTRAPOLATION **
C       ***********************************************
C
        CP(1,COUNTX)=SUMX*DX
C
C       ******************************************
C       ** RICHARDSON EXTRAPOLATION CALCULATION **
C       ******************************************
C
        C1=1.0
C
        DO Q=2,COUNTX
          C1=2*C1
          CP(Q,COUNTX)=(C1*CP(Q-1,COUNTX)-CP(Q-1,COUNTX-1))/(C1-1.0)
        ENDDO
C
        GI=CP(COUNTX,COUNTX)
C
        NEW=CP(COUNTX,COUNTX)
C
C       ***************************************
C       ** CALCULATE ERROR AFTER FIRST PASS **
C       ***************************************
C
        IF(COUNTX.GT.1) ERROR=ABS(OLD-NEW)/ABS(NEW)
C
C       ******************************************************
C       ** CHECK ERROR AFTER FIRST PASS TO SEE IF INTEGRATION **
C       ** IS WITHIN TOLERANCE END CALCULATION IF IT IS        **
C       ******************************************************
C
        IF((COUNTX.GT.1).AND.(ERROR.LT.TOL)) FINISHED=.TRUE.
C
C       ******************************************************
C       ** CHECK TO SEE IF WE HAVE REACHED THE MAXIMUM NUMBER **
```

```
C      ** OF SUBSECTIONS ALLOWED BY USER                    **
C      ****************************************************
C
       IF(COUNTX.EQ.NMAX) FINISHED=.TRUE.
C
       COUNTX=COUNTX+1
C
       OLD=NEW
C
C
       END DO
C
       GI=GI/(2.0*PI)*TRI(ZPRM,H,MID)
C
       RETURN
       END
CC
CC
C      *************************
C      ** TRIANGULAR FUNCTION **
C      *************************
C
       FUNCTION TRI (ZP,HH,MID)
       REAL TRI,ZP,HH,MID
       TRI=1.0-ABS(ZP-MID)/HH
       RETURN
       END
CC
C
C      ****************************************************************
C      ** THIS SUBROUTINE APPROXIMATES THE INTEGRAL OF A FUNCTION  **
C      ** USING MIDPOINT INTEGRATION WITH RICHARDSON EXTRAPOLATION **
C      ****************************************************************
C
       REAL FUNCTION MIDINT(F,A,B,ZM,AS,H)
C
       IMPLICIT NONE
       REAL CP(16,16),C1,SUMX,F,AS,H,MID
       REAL NSX,DX,A,B,ZM,MIDPNT,ERROR,OLD,NEW,TOL
       INTEGER COUNTX,I,Q,NMAX
       LOGICAL FINISHED
       EXTERNAL F
C
       FINISHED=.FALSE.
       COUNTX=1
       TOL=5.0E-4
       NMAX=16
C
C
C      ********************************************
C      ** CALCULATE MIDPOINT FOR TRIANGULAR FUNCTION **
C      ********************************************
       IF((A.EQ.0.0).AND.(B.EQ.H))THEN
       MID=0.0
       ELSE
```

```
      MID=(A+B)/2.0
      ENDIF
C
      DO WHILE(.NOT.FINISHED)
C
C      ***********************************
C      ** CALCULATE NUMBER OF SUBSECTIONS **
C      ***********************************
       NSX=2.0**COUNTX
C
C      *************************
C      ** CALCULATE STEP-SIZE **
C      *************************
       DX=(B-A)/NSX
C
C      ******************************************************
C      ** EVALUATE AND SUM FUNCTION VALUE AT MIDPOINTS **
C      ******************************************************
C
       SUMX=0.0
C
       DO I=1,NSX
         MIDPNT=A+FLOAT(I-1)*DX+DX/2.0
         SUMX=SUMX+F(ZM,MIDPNT,AS,H,MID)
       ENDDO
C
C
C      ****************************************************
C      ** ASSIGN SUMATION TO ARRAY FOR EXTRAPOLATION **
C      ****************************************************
       CP(1,COUNTX)=SUMX*DX
C
C      *******************************************
C      ** RICHARDSON EXTRAPOLATION CALCULATION **
C      *******************************************
C
       C1=1.0
C
       DO Q=2,COUNTX
         C1=2*C1
         CP(Q,COUNTX)=(C1*CP(Q-1,COUNTX)-CP(Q-1,COUNTX-1))/(C1-1.0)
       ENDDO
C
      MIDINT=CP(COUNTX,COUNTX)
C
      NEW=CP(COUNTX,COUNTX)
C
C      ***********************************
C      ** CALCULATE ERROR AFTER FIRST PASS **
C      ***********************************
C
      IF(COUNTX.GT.1) ERROR=ABS(OLD-NEW)/ABS(NEW)
C
C      ***********************************************************
C      ** CHECK ERROR AFTER FIRST PASS TO SEE IF INTEGRATION  **
```

```
C          ** IS WITHIN TOLERANCE END CALCULATION IF IT IS        **
C          ************************************************************
C
           IF((COUNTX.GT.1).AND.(ERROR.LT.TOL)) FINISHED=.TRUE.
C
C          ************************************************************
C          ** CHECK TO SEE IF WE HAVE REACHED THE MAXIMUM NUMBER **
C          ** OF SUBSECTIONS ALLOWED BY USER                      **
C          ************************************************************
C
           IF(COUNTX.EQ.NMAX) FINISHED=.TRUE.
C
           COUNTX=COUNTX+1
C
           OLD=NEW
C
C
        END DO
C
        RETURN
        END
```

# Appendix C:
# Chapter 4
# FORTRAN Computer Programs

```
C     ****************************************************************
C     ** THIS PROGRAM SOLVES FOR THE CURRENTS ON A RESISTIVE **
C     ** STRIP DUE TO A 1 V/M TM POLARIZED PLANE WAVE USING  **
C     ** THE MOMENT METHOD AND PULSE EXPANSIONS FUNCTIONS    **
C     ** WITH POINT MATCHING. THE EXACT KERNEL IS USED IN    **
C     ** THIS FORMULATION. THE MATCH POINT IS AT THE CENTER  **
C     ** OF EACH PULSE. THE FIELD FROM THE CURRENT IS USED   **
C     ** TO CALCULATE THE NORMALIZED RADAR CROSS SECTION     **
C     ** THIS PROGRAM IMPLEMENTS EQUATIONS (4.8a),(4.8b)     **
C     **                                                     **
C     **   RANDY BANCROFT  9/8/95                            **
C     **                                                     **
C     ****************************************************************
C     **                                                     **
C     ** VARIABLE DICTIONARY (MATRIX SECTION NOT INCLUDED)   **
C     **                 (INTEGRATION SECTIONS NOT INCLUDED) **
C     **                                                     **
C     ** A -- MOMENT MATRIX                                  **
C     ** B -- ENFORCEMENT MATRIX (BECOMES SOLUTION MATRIX)   **
C     ** C -- SPEED OF LIGHT IN FREE SPACE                   **
C     ** C1 -- CONSTANT USED IN RICHARDSON'S EXTRAPOLATION   **
C     ** CHARGE -- TOTAL CHARGE ON STRIP                     **
C     ** CP -- ARRAY FOR RICHARDSON'S EXTRAPOLATION          **
C     ** CSUM -- VARIABLE FOR COMPLEX SUMMATION              **
C     ** ELEMENT -- TEMPORARY VARIABLE FOR A(I,J)            **
C     ** ETA -- CHARACTERISTIC IMPEDANCE OF FREE SPACE       **
C     ** FREQ -- FREQUENCY OF PLANE WAVE                     **
C     ** G -- GREEN'S FUNCTION                               **
C     ** H -- SEGMENT LENGTH                                 **
C     ** I -- LOOP VARIABLE                                  **
C     ** INC -- INCIDENT ANGLE OF PLANE WAVE                 **
C     ** J -- LOOP VARIABLE                                  **
C     ** JAY -- SQUARE ROOT OF -1                            **
C     ** K -- FREE SPACE WAVENUMBER                          **
C     ** MTX -- ARRAY SIZE [A],[B]                           **
C     ** NMAX -- MAXIMUM NUMBER OF SEGMENT DIVISIONS         **
C     ** NS -- NUMBER OF SEGMENTS                            **
C     ** PI -- RATIO OF CIRCUMFERENCE TO DIAMETER OF CIRCLE  **
C     ** Q -- RICHARDSON'S EXTRAPOLATION LOOP VARIABLE       **
C     ** R -- SHEET RESISTANCE OF STRIP IN OHMS/SQUARE       **
C     ** S -- SEGMENT LOOP COUNTER                           **
C     ** SIGMA -- NORMALIZED RADAR CROSS SECTION             **
C     ** W -- WIDTH OF RESISTIVE STRIP                       **
C     ** XPRIME -- MATCH POINT DURING CURRENT INTEGRATION    **
C     ** ZM -- MATCH POINT                                   **
C     ** ZN -- LOWER LIMIT OF INTEGRATION OVER SEGMENT N     **
C     ** ZN1 -- UPPER LIMIT OF INTEGRATION OVER SEGMENT N    **
C     ** Y -- ARRAY WHICH HOLDS NUMBER OF SEGMENTS FOR S     **
C     ****************************************************************
C
C
      PROGRAM RSTRIPTM
C
      IMPLICIT NONE
```

```
C
      INTEGER MTX
      PARAMETER(MTX=64)
C
C     *****************************
C     ** DECLARE REAL VARIABLES **
C     *****************************
C
      REAL PI,C,ETA,ZN,ZN1,ZM,LAMBDA,K,C1,H,R
      REAL FREQ,SIGMA,CP(10,10),W,INC,XPRIME,MAX
C
C     ********************************
C     ** DECLARE COMPLEX VARIBLES **
C     ********************************
C
      COMPLEX JAY,A(MTX,MTX),B(MTX),ELEMENT,CSUM,EINC,ATEMP,BTEMP
      COMPLEX X(MTX),SUM,MJK
C
C     ******************************
C     ** DECLARE INTEGER VARIBLES **
C     ******************************
C
      INTEGER I,J,N,S,NMAX,Y(10),Q,NS
      INTEGER K2,ICOL,IROW,COL,XTMP,XINDX(MTX)
C
C     ********************************
C     ** ASSIGN FUNDAMENTAL CONSTANTS **
C     ********************************
C
      JAY=(0.0,1.0)
      PI=ACOS(-1.0)
      C=2.997925E8
      ETA=376.73
C
C
C     ****************************
C     ** INPUT DESIRED OPTIONS **
C     ****************************
C
      WRITE(*,'(5X,A)')'INPUT WIDTH OF RESISTIVE STRIP (LAMBDA): '
      READ(*,*)W
C
      WRITE(*,'(5X,A)')'INPUT SHEET RESISTANCE OF STRIP IN OHMS/SQ: '
      READ(*,*)R
C
      WRITE(*,'(5X,A)')'INPUT INCIDENT ANGLE IN DEGREES: '
      READ(*,*)INC
C
      INC=INC*PI/180.0
C
C     ***********************
C     ** USE 1 LAMDA VALUES **
C     ***********************
C
      FREQ=C
```

```
      LAMBDA=C/FREQ
      K=2.0*PI/LAMBDA
C
C     *************************
C     ** MAXIMUM VALUE OF S **
C     ** USED TO DETERMINE  **
C     ** NUMBER OF SEGMENTS **
C     *************************
C
      NMAX=6
C
C     *********************************
C     ** OUTER LOOP CONTROLS HOW MANY **
C     ** SEGMENT DIVISIONS OCCUR     **
C     *********************************
C
      DO S=1,NMAX
C
C        ***************************
C        ** CALCULATE THE NUMBER **
C        ** OF SEGMENTS FOR THIS **
C        ** LOOP ITERATION       **
C        ***************************
C
         NS=2**(S)
C
C        ***************************
C        ** CALCULATE STEP LENGTH **
C        ***************************
C
         H=W/FLOAT(NS)
C
C        *********************************
C        ** FILL THE [A] AND [B] MATRIX **
C        *********************************
C
         DO I=1,NS
           DO J=1,NS
C
C             ***************************
C             ** CALCULATE MATCH POINT **
C             ***************************
C
              ZM=FLOAT(I-1)*H+0.5*H
C
C             **********************************
C             ** CALCULATE ZN AND ZN+1 WHICH **
C             ** ARE UPPER AND LOWER LIMITS  **
C             ** OF CURRENT INTEGRATION      **
C             ** OVER THE PRESENT SEGMENT    **
C             **********************************
C
C
              ZN=FLOAT(J-1)*H
              ZN1=FLOAT(J)*H
```

```
C
C          ********************************
C          ** EVALUATE INTEGRATED GREEN'S **
C          ** FUNCTION AT UPPER AND LOWER **
C          ** LIMITS                      **
C          ********************************
C
           CALL INTEGRAL (ZN,ZN1,ZM,ELEMENT)
C
C          *********************************************
C          ** IF THE [A] MATRIX ELEMENT IS A          **
C          ** DIAGONAL TERM ADD ON SHEET RESISTANCE **
C          *********************************************
           IF(I.EQ.J)THEN
             A(I,J)=K*ETA*ELEMENT/4.0+R
           ELSE
             A(I,J)=K*ETA*ELEMENT/4.0
           ENDIF
C
C          ***********************
C          ** PRINT [A] MATRIX **
C          ***********************
C
C          *********************************************
C          ** PRINT THE CURRENT [A] MATRIX VALUE **
C          *********************************************
           WRITE (*,'(A1,4X,A,I3,I3,A,F15.4,2X,F15.4)')'+','A(',I,J,
    ;      ')=',A(I,J)
C
       ENDDO
C
C          ***********************
C          ** FILL THE B MATRIX **
C          ***********************
C
       B(I)=EINC(ZM,INC)
C
   ENDDO
C
C      ****************************************
C      ** SET DUMMY VARIBLES FOR MATRIX ROUTINE **
C      ****************************************
C
   N=NS
C
C  ****************************
C  ** SOLVE FOR STRIP CURRENTS **
C  ****************************
C
C
C  *************************************************************
C  ** GAUSSIAN ELIMINATION ALGORITHM WITH TOTAL PIVOTING **
C  *************************************************************
C
C  ***************************************
C  ** INITIALIZE SOLUTION INDEX MATRIX **
```

```
C      ************************************
C
       DO I=1,N
         XINDX(I)=I
       END DO
C
C
C      ************************************
C      ** OUTER LOOP CONTROLS ELIMINATION **
C      ************************************
C
       DO K2=1,N-1
C
C      ************************************
C      ** SEARCH FOR MAXIMUM VALUE IN ARRAY **
C      ************************************
C
       MAX=0.0
C
       DO I=K2,N
         DO J=K2,N
           IF(CABS(A(I,J)).GT.MAX)THEN
             MAX=A(I,J)
C            ************************************
C            ** KEEP INDICES OF MAXIMUM ELEMENT **
C            ************************************
             IROW=I
             ICOL=J
           ENDIF
         END DO
       END DO
C
C      ******************************************************
C      ** DETERMINE IF A ROW EXCHANGE OR COLUMN EXCHANGE **
C      ** IS REQUIRED TO BRING ELEMENT TO PIVOT          **
C      ******************************************************
C
C      ******************************************************
C      ** IF THE COLUMN INDEX AND THE ROW INDEX MATCH **
C      ** A ROW EXCHANGE WILL BRING THE MAXIMUM        **
C      ** ELEMENT TO THE PIVOT POINT                   **
C      ******************************************************
       IF(IROW.NE.K2)THEN
C         ******************************************************
C         ** EXCHANGE IROW (I.E. MAX ELEMENT ROW) WITH ROW K2 **
C         ******************************************************
          DO I=1,N
C            ******************************
C            ** FIRST SWAP THE [A] MATRIX **
C            ******************************
             ATEMP=A(IROW,I)
             A(IROW,I)=A(K2,I)
             A(K2,I)=ATEMP
          END DO
C         ******************************
```

```
C       ** THEN SWAP THE [B] MATRIX **
C       *****************************
        BTEMP=B(IROW)
        B(IROW)=B(K2)
        B(K2)=BTEMP
      ENDIF

C    ************************************************************
C    ** EXCHANGE ICOL (I.E. MAX ELEMENT COLUMN WITH COLUMN K2 **
C    ************************************************************
      IF(ICOL.NE.K2)THEN

C       DO I=1,N
C       *******************************
C       ** FIRST SWAP THE [A] MATRIX **
C       *******************************
        ATEMP=A(I,ICOL)
        A(I,ICOL)=A(I,K2)
        A(I,K2)=ATEMP
      END DO
C       ***********************************
C       ** THEN SWAP THE [XINDX] MATRIX  **
C       ** WHICH DOES THE SOLUTION ORDER **
C       ** BOOK KEEPING                  **
C       ***********************************
        XTMP=XINDX(ICOL)
        XINDX(ICOL)=XINDX(K2)
        XINDX(K2)=XTMP
C
      ENDIF
C
C
C    *************************************
C    ** NORMALIZE EACH ROW AND ELIMINATE **
C    *************************************
C
      IF(CABS(A(K2,K2)).EQ.0.0) WRITE(*,*)'ZERO PIVOT ENCOUNTERED: ERROR
     ;1'
C
      DO J=K2+1,N
        MJK=A(J,K2)/A(K2,K2)
        DO COL=K2,N
         A(J,COL)=A(J,COL)-MJK*A(K2,COL)
        END DO
         B(J)=B(J)-MJK*B(K2)
      END DO
C
      WRITE(*,'(A1,A,I3,A,I3,A,20X)')'+','Matrix Row ',K2,' of ',N,
     ;' Total Rows'
C
      ENDDO
C
C    *********************
C    ** BACKSUBSTITUTION **
C    *********************
```

```
C
      X(N)=B(N)/A(N,N)
C
      DO I=(N-1),1,-1
        SUM=(0.0,0.0)
        DO J=(I+1),N
          SUM=SUM+A(I,J)*X(J)
        END DO
        X(I)=(B(I)-SUM)/A(I,I)
      END DO
C
C     ************************************************************
C     ** USE THE INDEX ARRAY TO LOCATE PERMUTATED SOLUTIONS **
C     ************************************************************
C
      DO I=1,N
        DO J=1,N
          IF(XINDX(J).EQ.I) B(I)=X(J)
        END DO
      END DO
C
C
C     ********************************************
C     ** CACULATE SUM WHICH IS PROPORTIONAL   **
C     ** TO THE SCATTERED FIELD AT INCIDENT   **
C     ** ANGLE OF THE PLANE WAVE              **
C     ********************************************
C
      CSUM=(0.0,0.0)
      DO I=1,NS
        XPRIME=FLOAT(I-1)*H+0.5*H
        CSUM=CSUM+B(I)*CEXP(JAY*K*XPRIME*COS(INC))
      ENDDO
C
      CSUM=CSUM*H*(-K)*ETA/4.0
C
C     ****************************************
C     ** CALCULATE THE RADAR CROSS SECTION **
C     ****************************************
C
      SIGMA=REAL(CSUM*CONJG(CSUM))*2.0/PI
C
C     **********************************
C     ** ASSIGN RCS TO AN ARRAY FOR  **
C     ** USE WITH RICHARDSON         **
C     ** EXTRAPOLATION               **
C     **********************************
C
      CP(1,S)=SIGMA
C
C     ****************************************
C     ** SAVE VALUE OF NUMBER OF SEGMENTS **
C     ** TO [Y] ARRAY                     **
C     ****************************************
C
```

```
      Y(S)=NS
C
C     ********************************************
C     ** RICHARDSON EXTRAPOLATION CALCULATION **
C     ********************************************
C
      C1=1
      DO Q=2,S
        C1=2*C1
        CP(Q,S)=(C1*CP(Q-1,S)-CP(Q-1,S-1))/(C1-1)
      ENDDO
C
C     *************************************
C     ** PRINT OUT CALCULATED VALUES OF **
C     ** RADAR CROSS SECTION           **
C     *************************************
C
      IF (S.EQ.1) THEN
        DO I=1,25
          WRITE (*,*) ' '
        ENDDO
      END IF
C
      IF (S.EQ.1) THEN
        WRITE (*,'(12X,A)')'Resistive Strip Equation Pulses with Point
     ; Matching: '
C
        WRITE (*,'(12X,A,F7.2,A,2X,A,F7.2,A)')'Sheet Resistance=',R,
     ;' Ohms/sq','Width = ',W,' lambda'
C
        WRITE(*,'(12X,A,F7.2,A)')'Incident Angle of Plane Wave =',
C
     ;   INC*180.0/PI,' Degrees'
C
        WRITE(*,*)' '
      END IF
C
      IF (S.EQ.1) WRITE (*,'(A1,A)')'+',
     ;                '
C
      IF (S.EQ.1) THEN
C
        WRITE (*,'(8X,A,14X,A)')'M','(RCS)'
        WRITE (*,*)' '
C
      END IF
C
      IF (S.EQ.1) WRITE (*,'(16X,A,8X,A)')'Sigma','EXTRAPOLATION'

      IF (S.EQ.1) WRITE (*,*)' '
      IF (S.EQ.1) WRITE (*,*)' '
C
      WRITE (*,'(A1,A)')'+',
     ;                '
```

```
C
      WRITE (*,'(5X,I3,1X,F12.4,2X,F12.4)') Y(S),CP(1,S),CP(S,S)
C
      WRITE (*,*) ' '
C
      ENDDO
C
C
      END
CC
C     **********************************************
C     ** CALCULATE INTEGRAL OF GREEN'S FUNCTION    **
C     ** THE INTEGRATION IS FROM A TO B WITH ZM    **
C     ** THE MATCH POINT. INT IS THE RETURNED      **
C     ** INTEGRATION VALUE                         **
C     **********************************************
C
      SUBROUTINE INTEGRAL (A,B,ZM,INT)
C
      IMPLICIT NONE
C
      REAL A,B,ZM,GR,GI,RLINT,IMINT,MIDINT
C
      COMPLEX INT
C
      EXTERNAL GR,GI
C
C     ****************************************
C     ** CALL MIDPOINT INTEGRATION ROUTINE **
C     ** FOR REAL AND IMAGINARY PART OF    **
C     ** GREEN'S FUNCTION                  **
C     ****************************************
      RLINT=MIDINT(GR,A,B,ZM)
      IMINT=MIDINT(GI,A,B,ZM)
      INT=CMPLX(RLINT,IMINT)
C
      RETURN
      END
CC
C     ***********************************
C     ** REAL PART OF GREEN'S FUNCTION **
C     ***********************************
C
      REAL FUNCTION GR(Z,ZPRM)
      IMPLICIT NONE
      REAL PI,K,Z,ZPRM,J0
C
      PI=ACOS(-1.0)
      K=2.0*PI
C
      GR=J0(K*ABS(Z-ZPRM))
C
      RETURN
      END
CC
```

```
C      *****************************************
C      ** IMAGINARY PART OF GREEN'S FUNCTION **
C      *****************************************
C
       REAL FUNCTION GI(Z,ZPRM)
       IMPLICIT NONE
       REAL PI,K,Z,ZPRM,YO
C
       PI=ACOS(-1.0)
       K=2.0*PI
C
       GI=-YO(K*ABS(Z-ZPRM))
C
       RETURN
       END
CC
C      ****************************
C      ** INCIDENT E-FIELD VALUES **
C      ****************************
C
       COMPLEX FUNCTION EINC(X,INC)
C
       IMPLICIT NONE
C
       REAL X,INC,PI,K
C
       COMPLEX JAY
C
       JAY=(0.0,1.0)
C
       PI=ACOS(-1.0)
       K=2.0*PI
C
       EINC=-CEXP(JAY*K*X*COS(INC))
C
       RETURN
       END
C
C      ****************************************************************
C      ** THIS SUBROUTINE APPROXIMATES THE INTEGRAL OF A FUNCTION    **
C      ** USING MIDPOINT INTEGRATION WITH RICHARDSON EXTRAPOLATION **
C      ****************************************************************
C
       REAL FUNCTION MIDINT(F,A,B,ZM)
C
       IMPLICIT NONE
       REAL CP(16,16),C1,SUMX,F
       REAL NSX,DX,A,B,ZM,MIDPNT,ERROR,OLD,NEW,TOL
       INTEGER COUNTX,I,Q,NMAX
       LOGICAL FINISHED
       EXTERNAL F
C
       FINISHED=.FALSE.
       COUNTX=1
       TOL=5.0E-5
```

```
      NMAX=16
C
      DO WHILE(.NOT.FINISHED)
C
C       ***********************************
C       ** CALCULATE NUMBER OF SUBSECTIONS **
C       ***********************************
        NSX=2.0**COUNTX
C
C       *************************
C       ** CALCULATE STEP-SIZE **
C       *************************
        DX=(B-A)/NSX
C
C       ****************************************************
C       ** EVALUATE AND SUM FUNCTION VALUE AT MIDPOINTS **
C       ****************************************************
C
        SUMX=0.0
C
        DO I=1,NSX
          MIDPNT=A+FLOAT(I-1)*DX+DX/2.0
          SUMX=SUMX+F(ZM,MIDPNT)
        ENDDO
C
C
C       ***************************************************
C       ** ASSIGN SUMATION TO ARRAY FOR EXTRAPOLATION **
C       ***************************************************
        CP(1,COUNTX)=SUMX*DX
C
C       ******************************************
C       ** RICHARDSON EXTRAPOLATION CALCULATION **
C       ******************************************
C
        C1=1.0
C
        DO Q=2,COUNTX
          C1=2*C1
          CP(Q,COUNTX)=(C1*CP(Q-1,COUNTX)-CP(Q-1,COUNTX-1))/(C1-1.0)
        ENDDO
C
        MIDINT=CP(COUNTX,COUNTX)
C
        NEW=CP(COUNTX,COUNTX)
C
C       ***********************************
C       ** CALCULATE ERROR AFTER FIRST PASS **
C       ***********************************
C
        IF(COUNTX.GT.1) ERROR=ABS(OLD-NEW)/ABS(NEW)
C
C       **********************************************************
C       ** CHECK ERROR AFTER FIRST PASS TO SEE IF INTEGRATION  **
C       ** IS WITHIN TOLERANCE END CALCULATION IF IT IS        **
```

```
C    ***********************************************************
C
     IF((COUNTX.GT.1).AND.(ERROR.LT.TOL)) FINISHED=.TRUE.

C    ***********************************************************
C    ** CHECK TO SEE IF WE HAVE REACHED THE MAXIMUM NUMBER **
C    ** OF SUBSECTIONS ALLOWED BY USER                     **
C    ***********************************************************

     IF(COUNTX.EQ.NMAX) FINISHED=.TRUE.

     COUNTX=COUNTX+1

     OLD=NEW


     END DO

     RETURN
     END

     ***********************************
     ** BESSEL FUNCTION SUBROUTINE JO **
     ***********************************
     FUNCTION JO(X)
     IMPLICIT NONE
     INTEGER I
     REAL AK(7),BK(7),AA,BB,A,X,JO,PRODUCT

     DATA AK/0.79788456,-0.00000077,-0.00552740,
     ;      -0.00009512,0.00137237,-0.00072805,
     ;       0.00014476/

     DATA BK/-0.78539816,-0.04166397,-0.00003954,
     ;       0.00262573,-0.00054125,-0.00029333,
     ;       0.00013558/

     IF(ABS(X).LE.3.0)THEN
       A=(X/2.0)*(X/2.0)
       PRODUCT=1.0-A/49.0

       DO I=6,1,-1
          PRODUCT=1.0-A/FLOAT(I*I)*PRODUCT
       ENDDO

       JO=PRODUCT

     ELSE
C
       A=3.0/X
       AA=A*AK(7)
       BB=A*BK(7)
C
       DO I=6,1,-1
```

```
            AA=AK(I)+A*AA
            BB=BK(I)+A*BB
          ENDDO
C
          BB=BB+X
          JO=AA/SQRT(X)*COS(BB)

        ENDIF

        RETURN
        END

C       ***********************************
C       ** BESSEL FUNCTION SUBROUTINE YO **
C       ***********************************
        FUNCTION YO(X)
        IMPLICIT NONE
        INTEGER I
        REAL AT(7),AK(7),BK(7),BO,JO,GAMMA,PI,A,X,PRODUCT,AA,BB,YO
C
        DATA AK/0.79788456,-0.00000077,-0.00552740,
       ;        -0.00009512,0.00137237,-0.00072805,
       ;         0.00014476/
C
        DATA BK/-0.78539816,-0.04166397,-0.00003954,
       ;         0.00262573,-0.00054125,-0.00029333,
       ;         0.00013558/
C
        GAMMA=0.57721567
        PI=ACOS(-1.0)
        A=(X/2.0)*(X/2.0)

        IF(ABS(X).LE.3)THEN
          AT(1)=1.0

          DO I=2,7
            AT(I)=AT(I-1)+1.0/FLOAT(I)
          ENDDO

          BO=2.0*(GAMMA+LOG(X/2.0))*JO(X)

          PRODUCT=AT(6)-A*AT(7)/49.0

          DO I=6,2,-1
             PRODUCT=AT(I-1)-A/FLOAT(I*I)*PRODUCT
          ENDDO

          YO=(BO+ 2.0*A*PRODUCT)/PI

        ELSE
C
          A=3.0/X
          AA=A*AK(7)
          BB=A*BK(7)
C
```

```
      DO I=6,1,-1
        AA=AK(I)+A*AA
        BB=BK(I)+A*BB
      ENDDO
C
      BB=BB+X
      YO=AA/SQRT(X)*SIN(BB)
      ENDIF
C
      RETURN
      END
```

```
C   *************************************************************
C   ** THIS PROGRAM SOLVES FOR THE CURRENTS ON A METAL       **
C   ** STRIP DUE TO A 1 V/M TE POLARIZED PLANE WAVE USING    **
C   ** THE MOMENT METHOD AND PULSE EXPANSION FUNCTIONS       **
C   ** WITH POINT MATCHING. THE EXACT KERNEL IS USED IN      **
C   ** THIS FORMULATION. THE MATCH POINT IS AT THE CENTER    **
C   ** OF EACH PULSE. THE FIELD FROM THE CURRENT IS USED     **
C   ** TO CALCULATE THE NORMALIZED RADAR CROSS SECTION       **
C   ** THIS PROGRAM IMPLEMENTS EQUATIONS (4.21),(4.22)       **
C   **                                                       **
C   **   RANDY BANCROFT  9/6/95                              **
C   **                                                       **
C   *************************************************************
C   **                                                       **
C   ** VARIABLE DICTIONARY (MATRIX SECTION NOT INCLUDED)     **
C   **                  (INTEGRATION SECTIONS NOT INCLUDED)  **
C   **                                                       **
C   ** A -- MOMENT MATRIX                                    **
C   ** A1 -- VECTOR POTENTIAL TERM OF [A] MATRIX             **
C   ** Q1,Q2,Q3,Q4 -- SCALAR POTENTIAL TERMS OF [A] MATRIX   **
C   ** B -- ENFORCEMENT MATRIX (BECOMES SOLUTION MATRIX)     **
C   ** C -- SPEED OF LIGHT IN FREE SPACE                     **
C   ** C1 -- CONSTANT USED IN RICHARDSON'S EXTRAPOLATION     **
C   ** CP -- ARRAY FOR RICHARDSON'S EXTRAPOLATION            **
C   ** CSUM -- VARIABLE FOR COMPLEX SUMMATION                **
C   ** EO -- PERMITIVITTY OF FREE SPACE                      **
C   ** EINC -- INCIDENT ELECTRIC FIELD ON STRIP              **
C   ** ETA -- CHARACTERISTIC IMPEDANCE OF FREE SPACE         **
C   ** FREQ -- FREQUENCY OF PLANE WAVE                       **
C   ** HX -- SEGMENT LENGTH                                  **
C   ** I -- LOOP VARIABLE                                    **
C   ** INC -- INCIDENT ANGLE OF PLANE WAVE                   **
C   ** J -- LOOP VARIABLE                                    **
C   ** JAY -- SQUARE ROOT OF -1                              **
C   ** K -- FREE SPACE WAVENUMBER                            **
C   ** LAMBDA -- FREE SPACE WAVELENGTH                       **
C   ** MTX -- ARRAY SIZE [A],[B]                             **
C   ** NMAX -- MAXIMUM NUMBER OF SEGMENT DIVISIONS           **
C   ** NS -- NUMBER OF SEGMENTS                              **
C   ** OMEGA -- ANGULAR FREQUENCY OF PLANE WAVE              **
C   ** PI -- RATIO OF CIRCUMFERENCE TO DIAMETER OF CIRCLE    **
C   ** Q -- RICHARDSON'S EXTRAPOLATION LOOP VARIABLE         **
C   ** S -- SEGMENT LOOP COUNTER                             **
C   ** SX -- LENGTH OF STRIP                                 **
C   ** SIGMA -- NORMALIZED RADAR CROSS SECTION               **
C   ** UO -- PERMEABILITY OF FREE SPACE                      **
C   ** XM -- MATCH POINT                                     **
C   ** XMMNS -- MATCHPOINT MINUS ONE HALF SEGMENT            **
C   ** XMPLS -- MATCHPOINT PLUS ONE HALF SEGMENT             **
C   ** XN -- LOWER LIMIT OF INTEGRATION OVER SEGMENT N       **
C   ** XNP1 -- UPPER LIMIT OF INTEGRATION OVER SEGMENT N     **
C   ** XNM1 -- LOWER LIMIT MINUS ONE SEGMENT LENGTH          **
C   ** XNMNS -- LOWER LIMIT MINUS ONE HALF SEGMENT LENGTH    **
C   ** XNPLS -- LOWER LIMIT PLUS ONE HALF SEGMENT LENGTH     **
```

```
C      ** Y -- ARRAY WHICH HOLDS NUMBER OF SEGMENTS FOR S     **
C      **********************************************************
C

       PROGRAM STRIPTE
C

       IMPLICIT NONE
C

       INTEGER MTX
       PARAMETER(MTX=64)

C
C      ****************************
C      ** DECLARE REAL VARIABLES **
C      ****************************
C

       REAL PI,EO,UO,C,ETA,LAMBDA,K,C1,FREQ,OMEGA,SIGMA,CP(10,10),HX,INC
       REAL XPRIME,SX,MAX,XNMNS,XNPLS,XN,XNM1,XNP1,XMMNS,XMPLS,XM
C
C      ******************************
C      ** DECLARE COMPLEX VARIBLES **
C      ******************************
C

       COMPLEX JAY,A(MTX,MTX),B(MTX),CSUM,EINC,A1,Q1,Q2,Q3,Q4
       COMPLEX ATEMP,BTEMP,X(MTX),SUM,MJK
C
C      ******************************
C      ** DECLARE INTEGER VARIBLES **
C      ******************************
C

       INTEGER I,J,N,S,NMAX,Y(10),Q,NS
       INTEGER K2,ICOL,IROW,COL,XTMP,XINDX(MTX)
C
C      ********************************
C      ** ASSIGN FUNDAMENTAL CONSTANTS **
C      ********************************
C

       JAY=(0.0,1.0)
       PI=ACOS(-1.0)
       C=2.997925E8
       EO=8.854223E-12
       UO=1.256640E-6
       ETA=376.73
C
C      ***************************
C      ** INPUT DESIRED OPTIONS **
C      ***************************
C

       WRITE(*,'(5X,A)')'INPUT LENGTH OF STRIP IN WAVELENGTHS (LAMBDA): '
       READ(*,*)SX
C

       WRITE(*,'(5X,A)')'INPUT INCIDENT ANGLE IN DEGREES: '
       READ(*,*)INC
C

       INC=INC*PI/180.0
C
```

```
C      *******************************
C      ** USE 1 METER LAMBDA VALUES **
C      *******************************
C
       FREQ=C
       OMEGA=2.0*PI*FREQ
       LAMBDA=C/FREQ
       K=2.0*PI/LAMBDA
C
C      **************************
C      ** MAXIMUM VALUE OF S **
C      ** USED TO DETERMINE  **
C      ** NUMBER OF SEGMENTS **
C      **************************
C
       NMAX=6
C
C      **********************************
C      ** OUTER LOOP CONTROLS HOW MANY **
C      ** SEGMENT DIVISIONS OCCUR      **
C      **********************************
C
       DO S=1,NMAX
C
C        **************************
C        ** CALCULATE THE NUMBER **
C        ** OF SEGMENTS FOR THIS **
C        ** LOOP ITERATION       **
C        **************************
C
         NS=2**S
C
C        ***************************
C        ** CALCULATE STEP LENGTH **
C        ***************************
C
         HX=SX/FLOAT(NS+1)
C
C        *********************************
C        ** FILL THE [A] AND [B] MATRIX **
C        *********************************
C
         DO I=1,NS
           DO J=1,NS
C
C            **************************
C            ** CALCULATE MATCH POINT **
C            ** ALONG STRIP SURFACE   **
C            **************************
C
C            **********************
C            ** XM MINUS ONE-HALF **
C            **********************
             XMMNS=FLOAT(I-1)*HX+HX/2.0-SX/2.0
C
```

```
C      ***********************
C      ** XM PLUS ONE-HALF **
C      ***********************
       XMPLS=FLOAT(I)*HX+HX/2.0-SX/2.0
C
C      ********
C      ** XM **
C      ********
C      XM=FLOAT(I)*HX-SX/2.0
C
C      *****************************************
C      ** CALCULATE UPPER AND LOWER LIMITS  **
C      ** OF CURRENT + CHARGE INTEGRATION   **
C      ** FOR THE PRESENT SEGMENT           **
C      *****************************************
C
C      ***********************
C      ** XN MINUS ONE-HALF **
C      ***********************
C      XNMNS=FLOAT(J-1)*HX+HX/2-SX/2.0
C
C      ***********************
C      ** XN PLUS ONE-HALF **
C      ***********************
C      XNPLS=FLOAT(J)*HX+HX/2-SX/2.0
C
C      ********
C      ** XN **
C      ********
C      XN=FLOAT(J)*HX-SX/2.0
C
C      ******************
C      ** XN MINUS ONE **
C      ******************
C      XNM1=XN-HX
C
C      *****************
C      ** XN PLUS ONE **
C      *****************
C      XNP1=XN+HX
C
C      **********************************
C      ** EVALUATE INTEGRATED GREEN'S **
C      ** FUNCTION                    **
C      **********************************
C
C      **********************
C      ** VECTOR POTENTIAL **
C      **********************
C
       CALL INTEGRAL (XNMNS,XNPLS,XM,A1)
C
C      **********************
C      ** SCALAR POTENTIAL **
C      **********************
```

```
                CALL INTEGRAL (XNM1,XN,XMPLS,Q1)
                CALL INTEGRAL (XN,XNP1,XMPLS,Q2)
                CALL INTEGRAL (XNM1,XN,XMMNS,Q3)
                CALL INTEGRAL (XN,XNP1,XMMNS,Q4)
C
                A(I,J)=-OMEGA*UO/4.0*(A1)*HX
       ;               +(Q1-Q2-Q3+Q4)/(HX*4.0*OMEGA*EO)
C
C          **********************
C          ** PRINT [A] MATRIX **
C          **********************
C
C          *****************************************
C          ** PRINT THE CURRENT [A] MATRIX VALUE **
C          *****************************************
                WRITE (*,30) I,J,A(I,J)
   30           FORMAT ('+',4X,'A(',I3,I3,')=',F15.4,2X,F15.4)
C
            ENDDO
C
C          ***********************
C          ** FILL THE B MATRIX **
C          ***********************
C
            B(I)=EINC(XM,INC)*HX
C
        ENDDO
C
C       ********************************************
C       ** SET DUMMY VARIBLES FOR MATRIX ROUTINE **
C       ********************************************
C
        N=NS
C
C       *******************************
C       ** SOLVE FOR STRIP CURRENTS **
C       *******************************
C
C    *************************************************************
C    ** GAUSSIAN ELIMINATION ALGORITHM WITH TOTAL PIVOTING **
C    *************************************************************
C
C       **********************************
C       ** INITIALIZE SOLUTION INDEX MATRIX **
C       **********************************
C
     DO I=1,N
        XINDX(I)=I
     END DO
C
C
C       ***********************************
C       ** OUTER LOOP CONTROLS ELIMINATION **
C       ***********************************
C
```

```
      DO K2=1,N-1
C
C     ***********************************
C     ** SEARCH FOR MAXIMUM VALUE IN ARRAY **
C     ***********************************
C
      MAX=0.0
C
      DO I=K2,N
        DO J=K2,N
          IF(CABS(A(I,J)).GT.MAX)THEN
            MAX=A(I,J)
C           ***********************************
C           ** KEEP INDICES OF MAXIMUM ELEMENT **
C           ***********************************
            IROW=I
            ICOL=J
          ENDIF
        END DO
      END DO
C
C     ********************************************************
C     ** DETERMINE IF A ROW EXCHANGE OR COLUMN EXCHANGE **
C     ** IS REQUIRED TO BRING ELEMENT TO PIVOT          **
C     ********************************************************
C
C
C     ********************************************************
C     ** IF THE COLUMN INDEX AND THE ROW INDEX MATCH **
C     ** A ROW EXCHANGE WILL BRING THE MAXIMUM        **
C     ** ELEMENT TO THE PIVOT POINT                   **
C     ********************************************************
      IF(IROW.NE.K2)THEN
C       ********************************************************
C       ** EXCHANGE IROW (I.E. MAX ELEMENT ROW) WITH ROW K2 **
C       ********************************************************
        DO I=1,N
C         *******************************
C         ** FIRST SWAP THE [A] MATRIX **
C         *******************************
          ATEMP=A(IROW,I)
          A(IROW,I)=A(K2,I)
          A(K2,I)=ATEMP
        END DO
C         *******************************
C         ** THEN SWAP THE [B] MATRIX **
C         *******************************
        BTEMP=B(IROW)
        B(IROW)=B(K2)
        B(K2)=BTEMP
      ENDIF
C
C     ************************************************************
C     ** EXCHANGE ICOL (I.E. MAX ELEMENT COLUMN WITH COLUMN K2 **
C     ************************************************************
      IF(ICOL.NE.K2)THEN
```

```
C
        DO I=1,N
C           *******************************
C           ** FIRST SWAP THE [A] MATRIX **
C           *******************************
            ATEMP=A(I,ICOL)
            A(I,ICOL)=A(I,K2)
            A(I,K2)=ATEMP
        END DO
C           ***********************************
C           ** THEN SWAP THE [XINDX] MATRIX  **
C           ** WHICH DOES THE SOLUTION ORDER **
C           ** BOOK KEEPING                  **
C           ***********************************
            XTMP=XINDX(ICOL)
            XINDX(ICOL)=XINDX(K2)
            XINDX(K2)=XTMP
C
        ENDIF
C
C
C       **************************************
C       ** NORMALIZE EACH ROW AND ELIMINATE **
C       **************************************
C
        IF(CABS(A(K2,K2)).EQ.0.0) WRITE(*,*) 'ZERO PIVOT ENCOUNTERED: ERRO
       ;R 1'
C
        DO J=K2+1,N
          MJK=A(J,K2)/A(K2,K2)
          DO COL=K2,N
           A(J,COL)=A(J,COL)-MJK*A(K2,COL)
          END DO
          B(J)=B(J)-MJK*B(K2)
        END DO
C
        ENDDO
C
C       **********************
C       ** BACKSUBSTITUTION **
C       **********************
C
        X(N)=B(N)/A(N,N)
C
        DO I=(N-1),1,-1
          SUM=(0.0,0.0)
          DO J=(I+1),N
            SUM=SUM+A(I,J)*X(J)
          END DO
          X(I)=(B(I)-SUM)/A(I,I)
        END DO
C
C       ********************************************************
C       ** USE THE INDEX ARRAY TO LOCATE PERMUTATED SOLUTIONS **
C       ********************************************************
```

```
C
      DO I=1,N
        DO J=1,N
          IF(XINDX(J).EQ.I) B(I)=X(J)
        END DO
      END DO
C
C     ******************************************
C     ** CACULATE SUM WHICH IS PROPORTIONAL  **
C     ** TO THE SCATTERED FIELD AT INCIDENT   **
C     ** ANGLE OF THE PLANE WAVE             **
C     ******************************************
C
      CSUM=(0.0,0.0)
      DO I=1,NS
        XPRIME=FLOAT(I)*HX-SX
        CSUM=CSUM+B(I)*HX*CEXP(JAY*K*(XPRIME*COS(INC)))
      ENDDO
C
C     ******************************************
C     ** CALCULATE THE RADAR CROSS SECTION **
C     ******************************************
C
      SIGMA=REAL(CSUM*CONJG(CSUM))*K*ETA*ETA/4.0
C
C     ********************************
C     ** ASSIGN RCS TO AN ARRAY FOR **
C     ** USE WITH RICHARDSON        **
C     ** EXTRAPOLATION              **
C     ********************************
C
      CP(1,S)=ABS(SIGMA)
C
C     **************************************
C     ** SAVE VALUE OF NUMBER OF SEGMENTS **
C     ** TO [Y] ARRAY                     **
C     **************************************
C
      Y(S)=NS
C
C     *********************************************
C     ** RICHARDSON EXTRAPOLATION CALCULATION **
C     *********************************************
C
      C1=1
      DO Q=2,S
        C1=2*C1
        CP(Q,S)=(C1*CP(Q-1,S)-CP(Q-1,S-1))/(C1-1)
      ENDDO
C
C     ************************************
C     ** PRINT OUT CALCULATED VALUES OF **
C     ** RADAR CROSS SECTION            **
C     ************************************
C
```

```
      IF (S.EQ.1) THEN
        DO I=1,25
          WRITE (*,*) ' '
        ENDDO
      END IF
C
      IF (S.EQ.1) THEN
C
        WRITE (*,'(12X,A)')'Metallic Strip TE Polarization Pulses with
    ; Point Matching:  '

        WRITE (*,*) ' '
C
        WRITE (*,'(12X,A,F7.2,2X,A)')'Length of Strip =',SX,' lambda'
C
        WRITE(*,'(12X,A,F7.2,A)')'Incident Angle of Plane Wave =',

    ;   INC*180.0/PI,' degrees'

        WRITE(*,*)' '
C
      END IF
C
C
      IF (S.EQ.1) WRITE (*,'(A1,A)')'+',                '
    ;                                                   '
C
      IF (S.EQ.1)WRITE (*,'(8X,A,5X,A,13X,A)')'M','Sigma','EXTRAPOLATI
    ;ON'
C
      IF (S.EQ.1) WRITE (*,*) ' '
      IF (S.EQ.1) WRITE (*,*) ' '
      WRITE (*,'(A1,A)')'+',                '
    ;                                       '
      WRITE (*,'(A1,5X,I3,1X,F12.5,8X,F12.5)')'+',Y(S),CP(1,S),CP(S,S)
      WRITE (*,*) ' '
C
      ENDDO
C
      END
CC
C     **********************************************
C     ** CALCULATE INTEGRAL OF GREEN'S FUNCTION    **
C     ** THE INTEGRATION IS FROM A TO B WITH ZM    **
C     ** THE MATCH POINT. INT IS THE RETURNED      **
C     ** INTEGRATION VALUE                         **
C     **********************************************
C
      SUBROUTINE INTEGRAL (A,B,XM,INT)
C
      IMPLICIT NONE
C
      REAL A,B,GR,GI,RLINT,IMINT,MIDINT,XM
C
      COMPLEX INT
```

```
C
        EXTERNAL GR,GI
C
C       ***************************************
C       ** CALL MIDPOINT INTEGRATION ROUTINE **
C       ** FOR REAL AND IMAGINARY PART OF     **
C       ** GREEN'S FUNCTION                   **
C       ***************************************
        RLINT=MIDINT(GR,A,B,XM)
        IMINT=MIDINT(GI,A,B,XM)
        INT=CMPLX(RLINT,IMINT)
C
        RETURN
        END
CC
C       ***********************************
C       ** REAL PART OF GREEN'S FUNCTION **
C       ***********************************
C
        REAL FUNCTION GR(Z)
        IMPLICIT NONE
        REAL PI,K,Z,JO
C
        PI=ACOS(-1.0)
        K=2.0*PI
C
        GR=JO(K*ABS(Z))
C
        RETURN
        END
CC
C       ****************************************
C       ** IMAGINARY PART OF GREEN'S FUNCTION **
C       ****************************************
C
        REAL FUNCTION GI(Z)
        IMPLICIT NONE
        REAL PI,K,Z,YO
C
        PI=ACOS(-1.0)
        K=2.0*PI
C
        GI=-YO(K*ABS(Z))
C
        RETURN
        END
CC
C       ****************************
C       ** INCIDENT E-FIELD VALUES **
C       ****************************
C
        COMPLEX FUNCTION EINC(X,INC)
C
        IMPLICIT NONE
C
```

```
      REAL X,INC,PI,K
C
      COMPLEX JAY
C
      JAY=(0.0,1.0)
C
      PI=ACOS(-1.0)
      K=2.0*PI
C
      EINC=-SIN(INC)*CEXP(-JAY*K*X*COS(INC))
C
      RETURN
      END
C
C
C     ****************************************************************
C     ** THIS SUBROUTINE APPROXIMATES THE INTEGRAL OF A FUNCTION  **
C     ** USING MIDPOINT INTEGRATION WITH RICHARDSON EXTRAPOLATION **
C     ****************************************************************
C
      REAL FUNCTION MIDINT(F,A,B,XM)
C
      IMPLICIT NONE
      REAL CP(16,16),C1,SUMX,F
      REAL NSX,DX,A,B,MIDPNT,ERROR,OLD,NEW,TOL,XM
      INTEGER COUNTX,I,Q,NMAX
      LOGICAL FINISHED
      EXTERNAL F
C
      FINISHED=.FALSE.
      COUNTX=1
      TOL=5.0E-5
      NMAX=16
C
      DO WHILE(.NOT.FINISHED)
C
C        ************************************
C        ** CALCULATE NUMBER OF SUBSECTIONS **
C        ************************************
         NSX=2.0**COUNTX
C
C        **************************
C        ** CALCULATE STEP-SIZE **
C        **************************
         DX=(B-A)/NSX
C
C        *************************************************
C        ** EVALUATE AND SUM FUNCTION VALUE AT MIDPOINTS **
C        *************************************************
C
         SUMX=0.0
C
         DO I=1,NSX
           MIDPNT=A+FLOAT(I-1)*DX+DX/2.0
           SUMX=SUMX+F(ABS(XM-MIDPNT))
         ENDDO
```

```
C
C
C        ******************************************
C        ** ASSIGN SUMATION TO ARRAY FOR EXTRAPOLATION **
C        ******************************************
C        CP(1,COUNTX)=SUMX*DX
C
C        ******************************************
C        ** RICHARDSON EXTRAPOLATION CALCULATION **
C        ******************************************
C
C        C1=1.0
C
         DO Q=2,COUNTX
           C1=2*C1
           CP(Q,COUNTX)=(C1*CP(Q-1,COUNTX)-CP(Q-1,COUNTX-1))/(C1-1.0)
         ENDDO
C
         MIDINT=CP(COUNTX,COUNTX)
C
         NEW=CP(COUNTX,COUNTX)
C
C        ******************************************
C        ** CALCULATE ERROR AFTER FIRST PASS **
C        ******************************************
C
         IF(COUNTX.GT.1) ERROR=ABS(OLD-NEW)/ABS(NEW)
C
C        *************************************************************
C        ** CHECK ERROR AFTER FIRST PASS TO SEE IF INTEGRATION  **
C        ** IS WITHIN TOLERANCE END CALCULATION IF IT IS        **
C        *************************************************************
C
         IF((COUNTX.GT.1).AND.(ERROR.LT.TOL)) FINISHED=.TRUE.
C
C        *************************************************************
C        ** CHECK TO SEE IF WE HAVE REACHED THE MAXIMUM NUMBER **
C        ** OF SUBSECTIONS ALLOWED BY USER                     **
C        *************************************************************
C
         IF(COUNTX.EQ.NMAX) FINISHED=.TRUE.
C
         COUNTX=COUNTX+1
C
         OLD=NEW
C
C
         END DO
C
         RETURN
         END
C
C        ******************************************
C        ** BESSEL FUNCTION SUBROUTINE J0 **
```

```
C      **********************************
       FUNCTION JO(X)
       IMPLICIT NONE
       INTEGER I
       REAL AK(7),BK(7),AA,BB,A,X,JO,PRODUCT
C
       DATA AK/0.79788456,-0.00000077,-0.00552740,
     ;        -0.00009512,0.00137237,-0.00072805,
     ;         0.00014476/
C
       DATA BK/-0.78539816,-0.04166397,-0.00003954,
     ;         0.00262573,-0.00054125,-0.00029333,
     ;         0.00013558/
C
       IF(ABS(X).LE.3.0)THEN
         A=(X/2.0)*(X/2.0)
         PRODUCT=1.0-A/49.0

         DO I=6,1,-1
           PRODUCT=1.0-A/FLOAT(I*I)*PRODUCT
         ENDDO

         JO=PRODUCT

       ELSE
C
         A=3.0/X
         AA=A*AK(7)
         BB=A*BK(7)
C
         DO I=6,1,-1
           AA=AK(I)+A*AA
           BB=BK(I)+A*BB
         ENDDO
C
         BB=BB+X
         JO=AA/SQRT(X)*COS(BB)

       ENDIF

       RETURN
       END

C      **********************************
C      ** BESSEL FUNCTION SUBROUTINE YO **
C      **********************************
       FUNCTION YO(X)
       IMPLICIT NONE
       INTEGER I
       REAL AT(7),AK(7),BK(7),BO,JO,GAMMA,PI,A,X,PRODUCT,AA,BB,YO
C
       DATA AK/0.79788456,-0.00000077,-0.00552740,
     ;         -0.00009512,0.00137237,-0.00072805,
     ;          0.00014476/
C
```

```
      DATA BK/-0.78539816,-0.04166397,-0.00003954,
     ;           0.00262573,-0.00054125,-0.00029333,
     ;           0.00013558/
C
      GAMMA=0.57721567
      PI=ACOS(-1.0)
      A=(X/2.0)*(X/2.0)

      IF(ABS(X).LE.3)THEN
        AT(1)=1.0

        DO I=2,7
          AT(I)=AT(I-1)+1.0/FLOAT(I)
        ENDDO

        B0=2.0*(GAMMA+LOG(X/2.0))*J0(X)

        PRODUCT=AT(6)-A*AT(7)/49.0

        DO I=6,2,-1
           PRODUCT=AT(I-1)-A/FLOAT(I*I)*PRODUCT
        ENDDO

        YO=(B0+ 2.0*A*PRODUCT)/PI

      ELSE
C
        A=3.0/X
        AA=A*AK(7)
        BB=A*BK(7)
C
        DO I=6,1,-1
          AA=AK(I)+A*AA
          BB=BK(I)+A*BB
        ENDDO
C
        BB=BB+X
        YO=AA/SQRT(X)*SIN(BB)
      ENDIF
C
      RETURN
      END
```

# Appendix D:
# Chapter 5
# FORTRAN Computer Programs

```
C     ***********************************************************
C     ** THIS PROGRAM USES PULSE EXPANSION FUNCTIONS ALONG A **
C     ** CIRCULAR CONTOUR IN THE TM INTEGRAL EQUATION TO      **
C     ** SOLVE IT USING THE METHOD OF MOMENTS. THE MATCH      **
C     ** POINT IS AT THE CENTER OF EACH PULSE. THE FIELD      **
C     ** FROM THE CURRENT IS USED TO CALCULATE THE RADAR      **
C     ** CROSS SECTION (RCS) IN TWO DIMENSIONS                **
C     **                                                      **
C     ** THIS PROGRAM IMPLEMENTS (5.8) AND (5.9)              **
C     **                                                      **
C     ** RANDY BANCROFT  9/8/95                               **
C     **                                                      **
C     ***********************************************************
C     **                                                      **
C     ** VARIABLE DICTIONARY (MATRIX SECTION NOT INCLUDED)    **
C     **                  (INTEGRATION SECTIONS NOT INCLUDED) **
C     **                                                      **
C     ** A -- MOMENT MATRIX                                   **
C     ** AE -- RADIUS OF CIRCLE (ELLIPSE FORMULA)             **
C     ** B -- ENFORCEMENT MATRIX (BECOMES SOLUTION MATRIX)    **
C     ** BE -- RADIUS OF CIRCLE (ELLIPSE FORMULA)             **
C     ** C -- SPEED OF LIGHT IN FREE SPACE                    **
C     ** C1 -- CONSTANT USED IN RICHARDSON'S EXTRAPOLATION    **
C     ** CP -- ARRAY FOR RICHARDSON'S EXTRAPOLATION           **
C     ** CSUM -- VARIABLE FOR COMPLEX SUMMATION               **
C     ** EO -- PERMITIVITTY OF FREE SPACE                     **
C     ** EINC -- INCIDENT ELECTRIC FIELD ON CONTOUR           **
C     ** ELEMENT -- TEMPORARY VARIABLE FOR A(I,J)             **
C     ** ETA -- CHARACTERISTIC IMPEDANCE OF FREE SPACE        **
C     ** FREQ -- FREQUENCY OF PLANE WAVE                      **
C     ** H -- SEGMENT LENGTH                                  **
C     ** I -- LOOP VARIABLE                                   **
C     ** INC -- INCIDENT ANGLE OF PLANE WAVE                  **
C     ** J -- LOOP VARIABLE                                   **
C     ** JAY -- SQUARE ROOT OF -1                             **
C     ** K -- FREE SPACE WAVENUMBER                           **
C     ** L -- TOTAL ANGLE AROUND CONTOUR (I.E. 2*PI)          **
C     ** LAMBDA -- FREE SPACE WAVELENGTH                      **
C     ** MTX -- ARRAY SIZE [A],[B]                            **
C     ** NMAX -- MAXIMUM NUMBER OF SEGMENT DIVISIONS          **
C     ** NS -- NUMBER OF SEGMENTS                             **
C     ** OMEGA -- ANGULAR FREQUENCY OF PLANE WAVE             **
C     ** PI -- RATIO OF CIRCUMFERENCE TO DIAMETER OF CIRCLE   **
C     ** Q -- RICHARDSON'S EXTRAPOLATION LOOP VARIABLE        **
C     ** S -- SEGMENT LOOP COUNTER                            **
C     ** SIGMA -- NORMALIZED RADAR CROSS SECTION              **
C     ** THETA -- ANGLE OF CURRENT SEGMENT FOR INTEGRATION    **
C     **          (SAME AS MATCH POINT ANGLE)                 **
C     ** UO -- PERMEABILITY OF FREE SPACE                     **
C     ** ZM -- MATCH POINT                                    **
C     ** ZN -- LOWER LIMIT OF INTEGRATION OVER SEGMENT N      **
C     ** ZN1 -- UPPER LIMIT OF INTEGRATION OVER SEGMENT N     **
C     ** Y -- ARRAY WHICH HOLDS NUMBER OF SEGMENTS FOR S      **
C     ***********************************************************
```

```
      PROGRAM TMCIR

      IMPLICIT NONE

      INTEGER MTX
      PARAMETER(MTX=64)

      ****************************
      ** DECLARE REAL VARIABLES **
      ****************************

      REAL PI,C,ETA,ZN,ZN1,ZM,LAMBDA,K,L,C1,H,R,FREQ,SIGMA
      REAL CP(10,10),THETA,XR,AE,BE,INC,MAX

      ****************************
      ** DECLARE COMPLEX VARIBLES **
      ****************************

      COMPLEX JAY,A(MTX,MTX),B(MTX),ELEMENT,ATEMP,BTEMP,X(MTX),SUM,MJK
      COMPLEX CSUM,EINC

      ****************************
      ** DECLARE INTEGER VARIBLES **
      ****************************

      INTEGER I,J,N,S,NMAX,Y(10),Q,NS,K2,ICOL,IROW,COL,XTMP,XINDX(MTX)

      *********************************
      ** ASSIGN FUNDAMENTAL CONSTANTS **
      *********************************

      JAY=(0.0,1.0)
      PI=ACOS(-1.0)
      C=2.997925E8
      ETA=376.73


      ***************************
      ** INPUT DESIRED OPTIONS **
      ***************************

      WRITE(*,'(5X,A)')'INPUT DIAMETER OF CIRCULAR CONTOUR (LAMBDA): '
      READ(*,*)BE
      BE=BE/2.0
      AE=BE

      WRITE (*,'(5X,A)')'INPUT INCIDENT ANGLE IN DEGREES '
      READ (*,*)INC
      INC=INC*PI/180.0

      ***********************
      ** USE 1 LAMDA VALUES **
```

```
C     ************************
C
      FREQ=C
      LAMBDA=C/FREQ
      L=2.0*PI
      K=2.0*PI/LAMBDA
C
C     ************************
C     ** MAXIMUM VALUE OF S **
C     ** USED TO DETERMINE  **
C     ** NUMBER OF SEGMENTS **
C     ************************
C
      NMAX=5
C
C     **********************************
C     ** OUTER LOOP CONTROLS HOW MANY **
C     ** SEGMENT DIVISIONS OCCUR      **
C     **********************************
C
      DO S=1,NMAX
C
C        ************************
C        ** CALCULATE THE NUMBER **
C        ** OF SEGMENTS FOR THIS **
C        ** LOOP ITERATION       **
C        ************************
C
         NS=2**(S+1)
C
C        ************************
C        ** CALCULATE STEP LENGTH **
C        ************************
C
         H=L/FLOAT(NS)
C
C        ********************************
C        ** FILL THE [A] AND [B] MATRIX **
C        ********************************
C
         DO I=1,NS
           DO J=1,NS
C
C            ************************
C            ** CALCULATE MATCH POINT **
C            ************************
C
             ZM=FLOAT(I-1)*H+0.5*H
C
C            ********************************
C            ** CALCULATE ZN AND ZN+1 WHICH **
C            ** ARE UPPER AND LOWER LIMITS  **
C            ** OF CURRENT INTEGRATION      **
C            ** OVER THE PRESENT SEGMENT    **
C            ********************************
```

```
ZN=FLOAT(J-1)*H
ZN1=FLOAT(J)*H

*******************************
** EVALUATE INTEGRATED GREEN'S **
** FUNCTION AT UPPER AND LOWER **
** LIMITS                      **
*******************************

CALL INTEGRAL (ZN,ZN1,ZM,ELEMENT,AE,BE)
A(I,J)=K*ETA*ELEMENT/4.0

*********************
** PRINT [A] MATRIX **
*********************

*****************************************
** PRINT THE CURRENT [A] MATRIX VALUE **
*****************************************
WRITE (*,'(A1,4X,A,I3,I3,A,F15.4,2X,F15.4)')')'+','A(',I,J,')=',
A(I,J)

ENDDO

*********************
** FILL THE B MATRIX **
*********************

B(I)=EINC(ZM,AE,BE,INC)

ENDDO

*********************************************
** SET DUMMY VARIBLES FOR MATRIX ROUTINE **
*********************************************

N=NS

*******************************
** SOLVE FOR CONTOUR CURRENTS **
*******************************

*****************************************************************
** GAUSSIAN ELIMINATION ALGORITHM WITH TOTAL PIVOTING **
*****************************************************************

*************************************
** INITIALIZE SOLUTION INDEX MATRIX **
*************************************

DO I=1,N
  XINDX(I)=I
END DO
```

```
C
C
C          **********************************
C          ** OUTER LOOP CONTROLS ELIMINATION **
C          **********************************
C
           DO K2=1,N-1
C
C
C             **************************************
C             ** SEARCH FOR MAXIMUM VALUE IN ARRAY **
C             **************************************
C
              MAX=0.0
C
              DO I=K2,N
                DO J=K2,N
                  IF(CABS(A(I,J)).GT.MAX)THEN
                    MAX=A(I,J)
C                   **************************************
C                   ** KEEP INDICES OF MAXIMUM ELEMENT **
C                   **************************************
                    IROW=I
                    ICOL=J
                  ENDIF
                END DO
              END DO
C
C             ********************************************************
C             ** DETERMINE IF A ROW EXCHANGE OR COLUMN EXCHANGE **
C             ** IS REQUIRED TO BRING ELEMENT TO PIVOT          **
C             ********************************************************
C
C
C             *****************************************************
C             ** IF THE COLUMN INDEX AND THE ROW INDEX MATCH **
C             ** A ROW EXCHANGE WILL BRING THE MAXIMUM       **
C             ** ELEMENT TO THE PIVOT POINT                  **
C             *****************************************************
              IF(IROW.NE.K2)THEN
C             *****************************************************
C             ** EXCHANGE IROW (I.E. MAX ELEMENT ROW) WITH ROW K2 **
C             *****************************************************
                DO I=1,N
C                 *****************************
C                 ** FIRST SWAP THE [A] MATRIX **
C                 *****************************
                  ATEMP=A(IROW,I)
                  A(IROW,I)=A(K2,I)
                  A(K2,I)=ATEMP
                END DO
C               *****************************
C               ** THEN SWAP THE [B] MATRIX **
C               *****************************
                BTEMP=B(IROW)
                B(IROW)=B(K2)
                B(K2)=BTEMP
              ENDIF
```

```
C
C
C          *****************************************************************
C          ** EXCHANGE ICOL (I.E. MAX ELEMENT COLUMN WITH COLUMN K2 **
C          *****************************************************************
           IF(ICOL.NE.K2)THEN
C
             DO I=1,N
C               *******************************
C               ** FIRST SWAP THE [A] MATRIX **
C               *******************************
                ATEMP=A(I,ICOL)
                A(I,ICOL)=A(I,K2)
                A(I,K2)=ATEMP
             END DO
C            ***********************************
C            ** THEN SWAP THE [XINDX] MATRIX  **
C            ** WHICH DOES THE SOLUTION ORDER **
C            ** BOOK KEEPING                  **
C            ***********************************
             XTMP=XINDX(ICOL)
             XINDX(ICOL)=XINDX(K2)
             XINDX(K2)=XTMP
C
           ENDIF
C
C
C          *************************************
C          ** NORMALIZE EACH ROW AND ELIMINATE **
C          *************************************
C
           IF(CABS(A(K2,K2)).EQ.0.0) WRITE(*,*) 'ZERO PIVOT ENCOUNTERED:
      ;ERROR1'
C
           DO J=K2+1,N
             MJK=A(J,K2)/A(K2,K2)
             DO COL=K2,N
              A(J,COL)=A(J,COL)-MJK*A(K2,COL)
             END DO
              B(J)=B(J)-MJK*B(K2)
           END DO
C          ENDDO
C
C          *********************
C          ** BACKSUBSTITUTION **
C          *********************
C
           X(N)=B(N)/A(N,N)
C
           DO I=(N-1),1,-1
             SUM=(0.0,0.0)
             DO J=(I+1),N
               SUM=SUM+A(I,J)*X(J)
             END DO
             X(I)=(B(I)-SUM)/A(I,I)
```

```
      END DO
C
C     ***********************************************************
C     ** USE THE INDEX ARRAY TO LOCATE PERMUTATED SOLUTIONS **
C     ***********************************************************
C
      DO I=1,N
        DO J=1,N
          IF(XINDX(J).EQ.I) B(I)=X(J)
        END DO
      END DO
C
C     ***************************************************
C     ** CACULATE SUM WHICH IS PROPORTIONAL TO THE **
C     ** SCATTERED FIELD AT ZERO DEGREES INCIDENCE **
C     ***************************************************
C
      CSUM=(0.0,0.0)
      DO I=1,NS
        THETA=FLOAT(I-1)*H+0.5*H
        XR=COS(THETA)*R(THETA,AE,BE)
        CSUM=CSUM+B(I)*CEXP(-JAY*K*XR)*R(THETA,AE,BE)
      ENDDO
C
      CSUM=CSUM*H
C
C     *****************************************
C     ** CALCULATE THE RADAR CROSS SECTION **
C     *****************************************
C
      SIGMA=REAL(CSUM*CONJG(CSUM))*K*(ETA**2)/4.0
C
C     *************************************
C     ** ASSIGN RCS TO AN ARRAY FOR USE **
C     ** WITH RICHARDSON EXTRAPOLATION  **
C     *************************************
C
      CP(1,S)=SIGMA
C
C     **********************************************
C     ** SAVE NUMBER OF SEGMENTS TO [Y] ARRAY **
C     **********************************************
C
      Y(S)=NS
C
C     **********************************************
C     ** RICHARDSON EXTRAPOLATION CALCULATION **
C     **********************************************
C
      C1=1
      DO Q=2,S
        C1=2*C1
        CP(Q,S)=(C1*CP(Q-1,S)-CP(Q-1,S-1))/(C1-1)
      ENDDO
C
```

```
      IF (S.EQ.1) THEN
        DO I=1,25
          WRITE (*,*) ' '
        ENDDO
      END IF
C
      IF (S.EQ.1) THEN
        WRITE (*,*)' '
C
        WRITE (*,'(12X,A)')'TM Cylinder Eq.  Pulses with Point Matching
     ;:    '
C
        WRITE (*,'(12X,A,F7.3,A)')'DIAMETER=',2*BE,' WAVELENGTHS'
C
        WRITE (*,'(12X,A,F7.2,A)')'INCIDENT ANGLE = ',
     ;    180.0/PI*INC,' DEGREES'
C
        WRITE(*,*)' '
      END IF
C
      IF (S.EQ.1) WRITE (*,'(A1,A)')'+',' '
     ;                              '
C
      IF (S.EQ.1) THEN
        WRITE (*,'(8X,A,7X,A,6X,A)')'M','(RCS)','EXTRAPOLATION'
      END IF
C
      IF (S.EQ.1) WRITE (*,'(16X,A,9X,A)')'Sigma','Sigma'
C
      IF (S.EQ.1) WRITE (*,*) ' '
C
      WRITE (*,'(A1,A)')'+',' '
     ;                              '
C
      WRITE (*,'(A1,5X,I3,1X,F12.4,2X,F12.4)')'+',Y(S),CP(1,S),CP(S,S)
C
      WRITE(*,*)' '
C
      ENDDO
C
      END
C
C     **************************************************
C     ** CALCULATE INTEGRAL OF GREEN'S FUNCTION       **
C     ** THE INTEGRATION IS FROM A TO B WITH ZM       **
C     ** THE MATCH POINT. INT IS THE RETURNED         **
C     ** INTEGRATION VALUE                            **
C     **************************************************
C
      SUBROUTINE INTEGRAL (A,B,ZM,INT,AE,BE)
C
      IMPLICIT NONE
C
      REAL A,B,ZM,GR,GI,RLINT,IMINT,MIDINT,AE,BE
C
```

```
      COMPLEX INT
C
      EXTERNAL GR,GI
C
C     **************************************
C     ** CALL MIDPOINT INTEGRATION ROUTINE **
C     ** FOR REAL AND IMAGINARY PART OF    **
C     ** GREEN'S FUNCTION                  **
C     **************************************
      RLINT=MIDINT(GR,A,B,ZM,AE,BE)
      IMINT=MIDINT(GI,A,B,ZM,AE,BE)
      INT=CMPLX(RLINT,IMINT)
C
      RETURN
      END
CC
C     **************************************
C     ** REAL PART OF GREEN'S FUNCTION **
C     **************************************
C
      REAL FUNCTION GR(Z,ZPRM,AE,BE)
      IMPLICIT NONE
      REAL PI,K,Z,ZPRM,R,JO,DELTR,AE,BE
C
      PI=ACOS(-1.0)
      K=2.0*PI
C
      GR=JO(K*DELTR(Z,ZPRM,AE,BE))*R(ZPRM,AE,BE)
C
      RETURN
      END
CC
C     ****************************************
C     ** IMAGINARY PART OF GREEN'S FUNCTION **
C     ****************************************
C
      REAL FUNCTION GI(Z,ZPRM,AE,BE)
      IMPLICIT NONE
      REAL PI,K,Z,ZPRM,R,YO,DELTR,AE,BE
C
      PI=ACOS(-1.0)
      K=2.0*PI
C
      GI=-YO(K*DELTR(Z,ZPRM,AE,BE))*R(ZPRM,AE,BE)
C
      RETURN
      END
CC
C     ****************************
C     ** INCIDENT E-FIELD VALUES **
C     ****************************
C
      COMPLEX FUNCTION EINC(X,AE,BE,INC)
C
      IMPLICIT NONE
```

```
C
      REAL X,AE,BE,INC,PI,K,MAG,TOP,R
C
      COMPLEX JAY
C
      JAY=(0.0,1.0)
C
      PI=ACOS(-1.0)
      K=2.0*PI
C
      MAG=R(X,AE,BE)
C
      TOP=MAG*COS(X)*COS(INC)+MAG*SIN(X)*SIN(INC)
C
      EINC=COS(K*TOP)+JAY*SIN(K*TOP)
C
      RETURN
      END
C
CC
C
C     ***********************************************************
C     ** MAGNITUDE OF RADIUS VECTOR AS A FUNCTION OF THETA **
C     ***********************************************************
C
      REAL FUNCTION R(THETA,AE,BE)
C
      IMPLICIT NONE
C
      REAL THETA,MAG,AE,BE
C
      MAG=1.0/( (COS(THETA)/BE)**2 + (SIN(THETA)/AE)**2 )
C
      R=SQRT(MAG*(COS(THETA)**2+SIN(THETA)**2))
C
      RETURN
      END
C
C     ***********************************************************
C     ** MAGNITUDE OF THE DIFFERENCE OF TWO RADIUS VECTORS **
C     ***********************************************************
C
      REAL FUNCTION DELTR(THETA1,THETA2,AE,BE)
C
      IMPLICIT NONE
C
      REAL THETA1,THETA2,MAG1,MAG2,IHAT,JHAT,AE,BE,R
C
      MAG1=R(THETA1,AE,BE)
C
      MAG2=R(THETA2,AE,BE)
C
      IHAT=MAG1*COS(THETA1)-MAG2*COS(THETA2)
C
      JHAT=MAG1*SIN(THETA1)-MAG2*SIN(THETA2)
C
```

```
      DELTR=SQRT(IHAT**2+JHAT**2)
C
      RETURN
      END
C
C
C     ****************************************************************
C     ** THIS SUBROUTINE APPROXIMATES THE INTEGRAL OF A FUNCTION  **
C     ** USING MIDPOINT INTEGRATION WITH RICHARDSON EXTRAPOLATION **
C     ****************************************************************
C
      REAL FUNCTION MIDINT(F,A,B,ZM,AE,BE)
C
      IMPLICIT NONE
      REAL CP(16,16),C1,SUMX,F,AE,BE
      REAL NSX,DX,A,B,ZM,MIDPNT,ERROR,OLD,NEW,TOL
      INTEGER COUNTX,I,Q,NMAX
      LOGICAL FINISHED
      EXTERNAL F
C
      FINISHED=.FALSE.
      COUNTX=1
      TOL=1.0E-4
      NMAX=16
C
      DO WHILE(.NOT.FINISHED)
C
C        ************************************
C        ** CALCULATE NUMBER OF SUBSECTIONS **
C        ************************************
         NSX=2.0**COUNTX
C
C        **************************
C        ** CALCULATE STEP-SIZE **
C        **************************
         DX=(B-A)/NSX
C
C        ****************************************************
C        ** EVALUATE AND SUM FUNCTION VALUE AT MIDPOINTS **
C        ****************************************************
C
         SUMX=0.0
C
         DO I=1,NSX
           MIDPNT=A+FLOAT(I-1)*DX+DX/2.0
           SUMX=SUMX+F(ZM,MIDPNT,AE,BE)
         ENDDO
C
C
C        ************************************************
C        ** ASSIGN SUMATION TO ARRAY FOR EXTRAPOLATION **
C        ************************************************
         CP(1,COUNTX)=SUMX*DX
C
C        ********************************************
C        ** RICHARDSON EXTRAPOLATION CALCULATION **
```

```
C    ** SINGULARITY USE 2 OTHERWISE 4        **
C    *******************************************
C

     C1=1.0
C

     DO Q=2,COUNTX
       C1=2*C1
       CP(Q,COUNTX)=(C1*CP(Q-1,COUNTX)-CP(Q-1,COUNTX-1))/(C1-1.0)
     ENDDO
C

     MIDINT=CP(COUNTX,COUNTX)
C

     NEW=CP(COUNTX,COUNTX)
C
C    **************************************
C    ** CALCULATE ERROR AFTER FIRST PASS **
C    **************************************
C

     IF(COUNTX.GT.1) ERROR=ABS(OLD-NEW)/ABS(NEW)
C
C    ****************************************************************
C    ** CHECK ERROR AFTER FIRST PASS TO SEE IF INTEGRATION    **
C    ** IS WITHIN TOLERANCE END CALCULATION IF IT IS          **
C    ****************************************************************

     IF((COUNTX.GT.1).AND.(ERROR.LT.TOL)) FINISHED=.TRUE.

C    ****************************************************************
C    ** CHECK TO SEE IF WE HAVE REACHED THE MAXIMUM NUMBER **
C    ** OF SUBSECTIONS ALLOWED BY USER                     **
C    ****************************************************************

     IF(COUNTX.EQ.NMAX) FINISHED=.TRUE.

     COUNTX=COUNTX+1
     OLD=NEW

   END DO

   RETURN
   END
C

C    ***********************************
C    ** BESSEL FUNCTION SUBROUTINE JO **
C    ***********************************
     FUNCTION JO(X)
     IMPLICIT NONE
     INTEGER I
     REAL AK(7),BK(7),AA,BB,A,X,JO,PRODUCT

     DATA AK/0.79788456,-0.00000077,-0.00552740,
   ;        -0.00009512,0.00137237,-0.00072805,
   ;         0.00014476/
C

     DATA BK/-0.78539816,-0.04166397,-0.00003954,
```

```
;          0.00262573,-0.00054125,-0.00029333,
;          0.00013558/
C
    IF(ABS(X).LE.3.0)THEN
      A=(X/2.0)*(X/2.0)
      PRODUCT=1.0-A/49.0

      DO I=6,1,-1
         PRODUCT=1.0-A/FLOAT(I*I)*PRODUCT
      ENDDO

      JO=PRODUCT

    ELSE
C
      A=3.0/X
      AA=A*AK(7)
      BB=A*BK(7)
C
      DO I=6,1,-1
        AA=AK(I)+A*AA
        BB=BK(I)+A*BB
      ENDDO
C
      BB=BB+X
      JO=AA/SQRT(X)*COS(BB)

    ENDIF

    RETURN
    END

C   ***********************************
C   ** BESSEL FUNCTION SUBROUTINE YO **
C   ***********************************
    FUNCTION YO(X)
    IMPLICIT NONE
    INTEGER I
    REAL AT(7),AK(7),BK(7),BO,JO,GAMMA,PI,A,X,PRODUCT,AA,BB,YO
C
    DATA AK/0.79788456,-0.00000077,-0.00552740,
;          -0.00009512,0.00137237,-0.00072805,
;          0.00014476/
C
    DATA BK/-0.78539816,-0.04166397,-0.00003954,
;          0.00262573,-0.00054125,-0.00029333,
;          0.00013558/
C
    GAMMA=0.57721567
    PI=ACOS(-1.0)
    A=(X/2.0)*(X/2.0)

    IF(ABS(X).LE.3)THEN
      AT(1)=1.0
```

```
      DO I=2,7
        AT(I)=AT(I-1)+1.0/FLOAT(I)
      ENDDO

      BO=2.0*(GAMMA+LOG(X/2.0))*JO(X)

      PRODUCT=AT(6)-A*AT(7)/49.0

      DO I=6,2,-1
         PRODUCT=AT(I-1)-A/FLOAT(I*I)*PRODUCT
      ENDDO

      YO=(BO+ 2.0*A*PRODUCT)/PI

      ELSE
C
      A=3.0/X
      AA=A*AK(7)
      BB=A*BK(7)
C
      DO I=6,1,-1
        AA=AK(I)+A*AA
        BB=BK(I)+A*BB
      ENDDO
C
      BB=BB+X
      YO=AA/SQRT(X)*SIN(BB)
      ENDIF
C
      RETURN
      END
```

```
C     ***********************************************************
C     ** THIS PROGRAM USES PULSE EXPANSION FUNCTIONS ALONG A **
C     ** CIRCULAR CONTOUR IN THE TE INTEGRAL EQUATION TO      **
C     ** SOLVE IT USING THE METHOD OF MOMENTS. THE MATCH      **
C     ** POINT IS AT THE CENTER OF EACH PULSE. A NUMERICAL    **
C     ** DERIVATIVE IS TAKEN USING RICHARDSON'S EXTRA-        **
C     ** POLATION. THE FIELD FROM THE CURRENT IS USED TO      **
C     ** CALCULATE THE RADAR CROSS SECTION (RCS) IN TWO       **
C     ** DIMENSIONS                                           **
C     **                                                      **
C     ** THIS PROGRAM IMPLEMENTS (5.17) AND (5.18)            **
C     **                                                      **
C     ** RANDY BANCROFT  9/8/95                               **
C     **                                                      **
C     ***********************************************************
C     **                                                      **
C     ** VARIABLE DICTIONARY (MATRIX SECTION NOT INCLUDED)    **
C     **               (INTEGRATION SECTIONS NOT INCLUDED)    **
C     **                                                      **
C     ** A -- MOMENT MATRIX                                   **
C     ** AE -- RADIUS OF CIRCLE (ELLIPSE FORMULA)             **
C     ** B -- ENFORCEMENT MATRIX (BECOMES SOLUTION MATRIX)    **
C     ** BE -- RADIUS OF CIRCLE (ELLIPSE FORMULA)             **
C     ** C -- SPEED OF LIGHT IN FREE SPACE                    **
C     ** C1 -- CONSTANT USED IN RICHARDSON'S EXTRAPOLATION    **
C     ** CP -- ARRAY FOR RICHARDSON'S EXTRAPOLATION           **
C     ** CSUM -- VARIABLE FOR COMPLEX SUMMATION               **
C     ** EO -- PERMITIVITTY OF FREE SPACE                     **
C     ** ELEMENT -- TEMPORARY VARIABLE FOR A(I,J)             **
C     ** ETA -- CHARACTERISTIC IMPEDANCE OF FREE SPACE        **
C     ** FREQ -- FREQUENCY OF PLANE WAVE                      **
C     ** H -- SEGMENT LENGTH                                  **
C     ** HINC -- INCIDENT MAGNETIC FIELD ON CONTOUR           **
C     ** I -- LOOP VARIABLE                                   **
C     ** INC -- INCIDENT ANGLE OF PLANE WAVE                  **
C     ** J -- LOOP VARIABLE                                   **
C     ** JAY -- SQUARE ROOT OF -1                             **
C     ** K -- FREE SPACE WAVENUMBER                           **
C     ** L -- TOTAL ANGLE AROUND CONTOUR (I.E. 2*PI)          **
C     ** LAMBDA -- FREE SPACE WAVELENGTH                      **
C     ** MTX -- ARRAY SIZE [A],[B]                            **
C     ** NDOTR -- UNIT NORMAL DOTTED WITH RHAT                **
C     ** NMAX -- MAXIMUM NUMBER OF SEGMENT DIVISIONS          **
C     ** NS -- NUMBER OF SEGMENTS                             **
C     ** NX -- X COMPONENT OF UNIT NORMAL VECTOR AT           **
C     **        SEGMENT N                                     **
C     ** NY -- Y COMPONENT OF UNIT NORMAL VECTOR AT           **
C     **        SEGMENT N                                     **
C     ** OMEGA -- ANGULAR FREQUENCY OF PLANE WAVE             **
C     ** PI -- RATIO OF CIRCUMFERENCE TO DIAMETER OF CIRCLE   **
C     ** Q -- RICHARDSON'S EXTRAPOLATION LOOP VARIABLE        **
C     ** RHATX -- X COMPONENT OF UNIT VECTOR POINTING FROM    **
C     **          SOURCE POINT TO OBSERVATION POINT           **
C     ** RHATY -- Y COMPONENT OF UNIT VECTOR POINTING FROM    **
```

```
C    **               SOURCE POINT TO OBSERVATION POINT            **
C    ** RTOD -- CONVERSION FACTOR FOR RADIANS TO DEGREES           **
C    ** S -- SEGMENT LOOP COUNTER                                  **
C    ** SIGMA -- NORMALIZED RADAR CROSS SECTION                    **
C    ** THETA -- ANGLE OF CURRENT SEGMENT FOR INTEGRATION          **
C    **           (SAME AS MATCH POINT ANGLE)                      **
C    ** THETAB -- THETA BACK ONE SEGMENT                           **
C    ** THETAF -- THETA FORWARD ONE SEGMENT                        **
C    ** THETAN -- THETA AT CENTER OF N SEGMENT                     **
C    ** THETAM -- MATCH ANGLE                                      **
C    ** TX -- X COMPONENT OF TANGENT UNIT VECTOR                   **
C    **       TO SEGMENT N                                         **
C    ** TY -- Y COMPONENT OF TANGENT UNIT VECTOR                   **
C    **       TO SEGMENT N                                         **
C    ** UO -- PERMEABILITY OF FREE SPACE                           **
C    ** XPRIME -- X COORDINATE OF CURRENT                          **
C    ** Y -- ARRAY WHICH HOLDS NUMBER OF SEGMENTS FOR S            **
C    ** YPRIME -- Y COORDINATE OF CURRENT                          **
C    ** ZM -- MATCH POINT                                          **
C    ** ZN -- LOWER LIMIT OF INTEGRATION OVER SEGMENT N            **
C    ** ZN1 -- UPPER LIMIT OF INTEGRATION OVER SEGMENT N           **
C    ***************************************************************
C
     PROGRAM TECIR
C
     IMPLICIT NONE
C
     INTEGER MTX
     PARAMETER(MTX=64)
C
C    ****************************
C    ** DECLARE REAL VARIABLES **
C    ****************************
C
     REAL PI,C,ETA,ZN,ZN1,LAMBDA,K,L,C1,H,R,FREQ,SIGMA
     REAL CP(10,10),AE,BE,MAX,ZN,THETAM,THETAF,THETAN,THETAB
     REAL TX,TY,NX,NY,RHATX,RHATY,NDOTR,XPRIME,YPRIME,INC
C
C    *******************************
C    ** DECLARE COMPLEX VARIBLES **
C    *******************************
C
     COMPLEX JAY,A(MTX,MTX),B(MTX),CSUM,HINC,ELEMENT,ATEMP,BTEMP
     COMPLEX X(MTX),SUM,MJK
C
C    ******************************
C    ** DECLARE INTEGER VARIBLES **
C    ******************************
C
     INTEGER I,J,S,NMAX,Y(10),Q,WS,N,K2,ICOL,IROW,COL,XTMP,XINDX(MTX)
C
C    *********************************
C    ** ASSIGN FUNDAMENTAL CONSTANTS **
C    *********************************
C
```

```
        JAY=(0.0,1.0)
        PI=ACOS(-1.0)
        C=2.997925E8
        ETA=376.73
C
C       ***************************
C       ** INPUT DESIRED OPTIONS **
C       ***************************
C
        WRITE(*,'(5X,A)')'INPUT DIAMETER OF CIRCULAR CONTOUR (LAMBDA): '
        READ(*,*)BE
        BE=BE/2.0
        AE=BE
C
        WRITE(*,'(5X,A)')'INPUT INCIDENT ANGLE (DEGREES): '
        READ(*,*)INC
        INC=INC*PI/180.0
C
C       *************************
C       ** USE 1 LAMDA VALUES **
C       *************************
C
        FREQ=C
        LAMBDA=C/FREQ
        L=2.0*PI
        K=2.0*PI/LAMBDA
C
C       *************************
C       ** MAXIMUM VALUE OF S **
C       ** USED TO DETERMINE   **
C       ** NUMBER OF SEGMENTS  **
C       *************************
C
        NMAX=5
C
C       **********************************
C       ** OUTER LOOP CONTROLS HOW MANY **
C       ** SEGMENT DIVISIONS OCCUR      **
C       **********************************
C
        DO S=1,NMAX
C
C       ***************************
C       ** CALCULATE THE NUMBER **
C       ** OF SEGMENTS FOR THIS **
C       ** LOOP ITERATION       **
C       ***************************
C
          NS=2**(S+1)
C
C       ***************************
C       ** CALCULATE STEP LENGTH **
C       ***************************
C
          H=L/FLOAT(NS)
```

```
C
C
C       ********************************
C       ** FILL THE [A] AND [B] MATRIX **
C       ********************************
C
        DO I=1,NS
          DO J=1,NS
C
C       ***************************
C       ** CALCULATE MATCH POINT  **
C       ***************************
C
            ZM=FLOAT(I-1)*H+H/2.0
C
C       ********************************
C       ** CALCULATE ZN AND ZN+1 WHICH **
C       ** ARE UPPER AND LOWER LIMITS  **
C       ** OF CURRENT INTEGRATION      **
C       ** OVER THE PRESENT SEGMENT    **
C       ********************************
C
            ZN=FLOAT(J-1)*H
            ZN1=FLOAT(J)*H
C
C       ********************************
C       ** EVALUATE INTEGRATED GREEN'S **
C       ** FUNCTION AT UPPER AND LOWER **
C       ** LIMITS                      **
C       ********************************
C
            CALL INTEGRAL (ZN,ZN1,ZM,ELEMENT,AE,BE,H)
            ELEMENT=ELEMENT/(4.0*JAY)
C
            IF(I.EQ.J)THEN
              A(I,J)=ELEMENT+(0.5,0.0)
            ELSE
              A(I,J)=ELEMENT
            ENDIF
C
C       *********************
C       ** PRINT [A] MATRIX **
C       *********************
C
C       *******************************************
C       ** PRINT THE CURRENT [A] MATRIX VALUE **
C       *******************************************
        WRITE (*,'(A1,4X,A,I3,I3,A,F15.4,2X,F15.4)')'+','A(',I,J,')=',
     ;  A(I,J)
C
          ENDDO
C
C       *********************
C       ** FILL THE B MATRIX **
C
```

```
C         ***********************
C
          B(I)=-HINC(ZM,AE,BE,INC)
C
      ENDDO
C
C         *******************************************
C         ** SET DUMMY VARIBLES FOR MATRIX ROUTINE **
C         *******************************************
C
      N=NS
C
C         ********************************
C         ** SOLVE FOR CONTOUR CURRENTS **
C         ********************************
C
C         ****************************************************************
C         ** GAUSSIAN ELIMINATION ALGORITHM WITH TOTAL PIVOTING **
C         ****************************************************************
C
C         **************************************
C         ** INITIALIZE SOLUTION INDEX MATRIX **
C         **************************************
C
      DO I=1,N
        XINDX(I)=I
      END DO
C
C
C         ****************************************
C         ** OUTER LOOP CONTROLS ELIMINATION **
C         ****************************************
C
      DO K2=1,N-1
C
C            *****************************************
C            ** SEARCH FOR MAXIMUM VALUE IN ARRAY **
C            *****************************************
C
         MAX=0.0
C
         DO I=K2,N
           DO J=K2,N
             IF(CABS(A(I,J)).GT.MAX)THEN
               MAX=A(I,J)
C            ***********************************
C            ** KEEP INDICES OF MAXIMUM ELEMENT **
C            ***********************************
               IROW=I
               ICOL=J
             ENDIF
           END DO
         END DO
C
C         *****************************************************
```

```
C       ** DETERMINE IF A ROW EXCHANGE OR COLUMN EXCHANGE **
C       ** IS REQUIRED TO BRING ELEMENT TO PIVOT           **
C       ****************************************************
C
C       ****************************************************
C       ** IF THE COLUMN INDEX AND THE ROW INDEX MATCH **
C       ** A ROW EXCHANGE WILL BRING THE MAXIMUM        **
C       ** ELEMENT TO THE PIVOT POINT                   **
C       ****************************************************
        IF(IROW.NE.K2)THEN
C          **********************************************************
C          ** EXCHANGE IROW (I.E. MAX ELEMENT ROW) WITH ROW K2 **
C          **********************************************************
           DO I=1,N
C             ****************************
C             ** FIRST SWAP THE [A] MATRIX **
C             ****************************
              ATEMP=A(IROW,I)
              A(IROW,I)=A(K2,I)
              A(K2,I)=ATEMP
           END DO
C          ****************************
C          ** THEN SWAP THE [B] MATRIX **
C          ****************************
           BTEMP=B(IROW)
           B(IROW)=B(K2)
           B(K2)=BTEMP
C
        ENDIF
C
C
C       ****************************************************************
C       ** EXCHANGE ICOL (I.E. MAX ELEMENT COLUMN WITH COLUMN K2 **
C       ****************************************************************
        IF(ICOL.NE.K2)THEN
C
           DO I=1,N
C             ****************************
C             ** FIRST SWAP THE [A] MATRIX **
C             ****************************
              ATEMP=A(I,ICOL)
              A(I,ICOL)=A(I,K2)
              A(I,K2)=ATEMP
           END DO
C          ****************************
C          ** THEN SWAP THE [XINDX] MATRIX **
C          ** WHICH DOES THE SOLUTION ORDER **
C          ** BOOK KEEPING                   **
C          ****************************
           XTMP=XINDX(ICOL)
           XINDX(ICOL)=XINDX(K2)
           XINDX(K2)=XTMP
C
        ENDIF
C
C
```

```
C        ****************************************
C        ** NORMALIZE EACH ROW AND ELIMINATE **
C        ****************************************
C
         IF(CABS(A(K2,K2)).EQ.0.0) WRITE(*,*) 'ZERO PIVOT ENCOUNTERED: ER
      ;ROR1'
C
         DO J=K2+1,N
           MJK=A(J,K2)/A(K2,K2)
           DO COL=K2,N
            A(J,COL)=A(J,COL)-MJK*A(K2,COL)
           END DO
            B(J)=B(J)-MJK*B(K2)
         END DO
C
C
         WRITE(*,'(A1,A,I3,A,I3,A)')')'+','Matrix Row ',K2,' of ',N,
      ;  ' Rows              '
C
C
         ENDDO
C
C        **********************
C        ** BACKSUBSTITUTION **
C        **********************
C
         X(N)=B(N)/A(N,N)
C
         DO I=(N-1),1,-1
           SUM=(0.0,0.0)
           DO J=(I+1),N
             SUM=SUM+A(I,J)*X(J)
           END DO
           X(I)=(B(I)-SUM)/A(I,I)
         END DO
C
C        ************************************************************
C        ** USE THE INDEX ARRAY TO LOCATE PERMUTATED SOLUTIONS **
C        ************************************************************
C
         DO I=1,N
           DO J=1,N
             IF(XINDX(J).EQ.I) B(I)=X(J)
           END DO
         END DO
C
C        ***************************************************
C        ** CACULATE SUM WHICH IS PROPORTIONAL TO THE **
C        ** SCATTERED FIELD AT ZERO DEGREE INCIDENCE  **
C        ***************************************************
C
         CSUM=(0.0,0.0)
         DO I=1,NS
C
C        *************************
```

```
C       ** CALCULATE MATCH ANGLE **
C       **************************
C
        THETAM=INC
C
C       ********************************
C       ** CALCULATE UPPER AND LOWER   **
C       ** ANGLE LIMITS OF INTEGRATION **
C       ** OVER THE PRESENT SEGMENT    **
C       ********************************
C
C
        THETAF=FLOAT(I)*H
        THETAN=FLOAT(I-1)*H+0.5*H
        THETAB=FLOAT(I-1)*H
C
C       *************************************************************
C       ** CALCULATE APPROXIMATE TANGENT VECTOR COMPONENTS **
C       *************************************************************
        TX=R(THETAF,AE,BE)*COS(THETAF)-R(THETAB,AE,BE)*COS(THETAB)
        TY=R(THETAF,AE,BE)*SIN(THETAF)-R(THETAB,AE,BE)*SIN(THETAB)
C
C       ************************************************************
C       ** CALCULATE NORMAL VECTOR WHICH IS T CROSSED WITH Z **
C       ** DIVIDED BY ITS MAGNITUDE TO CREATE UNIT VECTOR    **
C       ************************************************************
        NX=TY/SQRT(TX**2+TY**2)
        NY=-TX/SQRT(TX**2+TY**2)
C
C       *****************************************
C       ** CALCULATE UNIT VECTOR POINTING FROM **
C       ** SOURCE TO FIELD POINT              **
C       *****************************************
C
        RHATX=COS(THETAM)
        RHATY=SIN(THETAM)
C
C       *****************************************
C       ** DOT RHAT WITH UNIT NORMAL VECTOR **
C       *****************************************
        NDOTR=NX*RHATX+NY*RHATY
C
        XPRIME=R(THETAN,AE,BE)*COS(THETAN)
        YPRIME=R(THETAN,AE,BE)*SIN(THETAN)
C
        CSUM=CSUM+B(I)*NDOTR*CEXP(JAY*K
     ;*(XPRIME*COS(THETAM)+YPRIME*SIN(THETAM)))*R(THETAN,AE,BE)

        ENDDO
C
        CSUM=CSUM*H
C
C       *****************************************
C       ** CALCULATE THE RADAR CROSS SECTION **
C       *****************************************
```

```
C
          SIGMA=REAL(CSUM*CONJG(CSUM))*ETA*ETA*K/4.0
C
C         ********************************
C         ** ASSIGN RCS TO AN ARRAY FOR  **
C         ** USE WITH RICHARDSON         **
C         ** EXTRAPOLATION               **
C         ********************************
C
          CP(1,S)=SIGMA
C
C         ************************************
C         ** SAVE VALUE OF NUMBER OF SEGMENTS **
C         ** TO [Y] ARRAY                **
C         ************************************
C
          Y(S)=NS
C
C         *********************************************
C         ** RICHARDSON EXTRAPOLATION CALCULATION **
C         *********************************************
C
          C1=1
          DO Q=2,S
            C1=2*C1
            CP(Q,S)=(C1*CP(Q-1,S)-CP(Q-1,S-1))/(C1-1)
          ENDDO
C
C         ************************************
C         ** PRINT OUT CALCULATED VALUES OF **
C         ** RADAR CROSS SECTION         **
C         ************************************
C
C
          IF (S.EQ.1) THEN
            DO I=1,25
              WRITE (*,*) ' '
            ENDDO
          END IF
C
          IF (S.EQ.1) THEN
            WRITE (*,'(12X,A)')'Harrington TE Cylinder Eq.  Pulses with Poi
         ;nt Matching:  '

            WRITE (*,*) ' '
C
            WRITE (*,'(12X,A,F7.3,A)')'Radius=',2*BE,' Wavelengths'

            WRITE(*,*)' '
          END IF
C
C
          IF (S.EQ.1) WRITE (*,'(A1,A)')'+',' '
        ;
C
```

```
      IF (S.EQ.1) THEN
C
        WRITE (*,'(8X,A,6X,A)')'M',' (RCS)         EXTRAPOLATION'
        WRITE(*,*)' '
C
      END IF
C
      IF (S.EQ.1) WRITE (*,'(16X,A,9X,A)')'Sigma','Sigma'

      IF (S.EQ.1) WRITE (*,*) ' '
      IF (S.EQ.1) WRITE (*,*) ' '
      WRITE (*,'(A1,A)')'+','

C
      WRITE (*,'(A1,5X,I3,1X,F12.4,2X,F12.4)')'+',Y(S),CP(1,S),CP(S,S)
C
      WRITE (*,*)
      WRITE (*,*)
C
      ENDDO
C
      END
CC
C     ****************************************************
C     ** CALCULATE INTEGRAL OF GREEN'S FUNCTION      **
C     ** THE INTEGRATION IS FROM A TO B WITH ZM      **
C     ** THE MATCH POINT. INT IS THE RETURNED        **
C     ** INTEGRATION VALUE                           **
C     ****************************************************
C
      SUBROUTINE INTEGRAL (A,B,ZM,INT,AE,BE,H)
C
      IMPLICIT NONE
C
      REAL A,B,ZM,GR,GI,RLINT,IMINT,MIDINT,AE,BE,H
C
      COMPLEX INT
C
      EXTERNAL GR,GI
C
C     ****************************************
C     ** CALL MIDPOINT INTEGRATION ROUTINE **
C     ** FOR REAL AND IMAGINARY PART OF    **
C     ** GREEN'S FUNCTION                  **
C     ****************************************
      RLINT=MIDINT(GR,A,B,ZM,AE,BE,H)
      IMINT=MIDINT(GI,A,B,ZM,AE,BE,H)
      INT=CMPLX(RLINT,IMINT)
C
      RETURN
      END
CC
C     **********************************
C     ** REAL PART OF GREEN'S FUNCTION **
C     **********************************
```

```
C
      REAL FUNCTION GR(Z,ZPRM,AE,BE,H)
      IMPLICIT NONE
      REAL PI,K,Z,ZPRM,R,JO,AE,BE,H,RX,RY,RCX,RCY
      REAL ZF,ZB,TX,TY,NX,NY,HX,HY,RPHX,RPHY,RMHX,RMHY,MPLUS,MMINUS
      REAL FPLUS,FMINUS,DGDN,C1,CP(10,10),NEW,OLD,TOL,ERROR
      INTEGER I,NS,Q,NMAX
      LOGICAL FINISHED
C
      PI=ACOS(-1.0)
      K=2.0*PI
      FINISHED=.FALSE.
      TOL=1.0E-5
      NMAX=8
C
C     *********************************************
C     ** CALCULATE X AND Y COMPONENTS OF RPRIME **
C     *********************************************
      RX=R(ZPRM,AE,BE)*COS(ZPRM)
      RY=R(ZPRM,AE,BE)*SIN(ZPRM)
C
C     *****************************************************************
C     ** CALCULATE X AND Y COMPONENTS OF RC (MATCH POINT ANGLE) **
C     *****************************************************************
      RCX=R(Z,AE,BE)*COS(Z)
      RCY=R(Z,AE,BE)*SIN(Z)
C
C     *********************************************
C     ** CALCULATE BOUNDING ANGLES OF PRESENT SEGMENT **
C     ** TO FIND APPROXIMATE TANGENT VECTOR          **
C     *********************************************
      ZF=ZPRM+0.5*H
      ZB=ZPRM-0.5*H
C
C     *************************************************************
C     ** CALCULATE APPROXIMATE TANGENT VECTOR COMPONENTS **
C     *************************************************************
      TX=R(ZF,AE,BE)*COS(ZF)-R(ZB,AE,BE)*COS(ZB)
      TY=R(ZF,AE,BE)*SIN(ZF)-R(ZB,AE,BE)*SIN(ZB)
C
C     *************************************************************
C     ** CALCULATE NORMAL VECTOR WHICH IS T CROSSED WITH Z **
C     ** DIVIDED BY ITS MAGNITUDE TO CREATE UNIT VECTOR   **
C     *************************************************************
      NX=TY/SQRT(TX**2+TY**2)
      NY=-TX/SQRT(TX**2+TY**2)
C
C     *****************************************************************
C     ** APPROXIMATE DERIVATIVE OF GREEN'S FUNCTION OF SOURCE POINT **
C     ** WITH RESPECT TO NORMAL DIRECTION TO CONTOUR              **
C     *****************************************************************
C
```

```
      I=1
C
      DO WHILE(.NOT.FINISHED)
C
      NS=2**I
C
      HX=NX/FLOAT(NS)
      HY=NY/FLOAT(NS)
C
C
C     *********************************************
C     ** CALCULATE RPRIME + SCALED NORMAL VECTOR **
C     *********************************************
C
      RPHX = RX+HX
      RPHY = RY+HY
C
C
C     *********************************************
C     ** CALCULATE RPRIME - SCALED NORMAL VECTOR **
C     *********************************************
      RMHX = RX-HX
      RMHY = RY-HY
C
C
C     ***************************************
C     ** CALCULATE APPROXIMATE DERIVATIVE **
C     ***************************************
      MPLUS=SQRT((RCX-RMHX)**2+(RCY-RMHY)**2)
      MMINUS=SQRT((RCX-RPHX)**2+(RCY-RPHY)**2)
C
      FPLUS=J0(K*(MPLUS))
      FMINUS=J0(K*(MMINUS))
C
      DGDN=(FPLUS-FMINUS)/(2.0*SQRT(HX*HX+HY*HY))
C
C
C
C
C     ************************************************
C     ** ASSIGN SUMATION TO ARRAY FOR EXTRAPOLATION **
C     ************************************************
      CP(1,I)=DGDN
C
C
C     *******************************************
C     ** RICHARDSON EXTRAPOLATION CALCULATION **
C     *******************************************
C
      C1=1.0
C
      DO Q=2,I
        C1=2*C1
        CP(Q,I)=(C1*CP(Q-1,I)-CP(Q-1,I-1))/(C1-1.0)
      ENDDO
C
      DGDN=CP(I,I)
C
      NEW=CP(I,I)
```

```
C
C
C          *************************************
C          ** CALCULATE ERROR AFTER FIRST PASS **
C          *************************************
C
           IF((I.GT.1).AND.(ABS(NEW).GT.0.0)) ERROR=ABS(OLD-NEW)/ABS(NEW)
C
C          ***********************************************************
C          ** CHECK ERROR AFTER FIRST PASS TO SEE IF INTEGRATION   **
C          ** IS WITHIN TOLERANCE END CALCULATION IF IT IS         **
C          ***********************************************************
C
           IF((I.GT.1).AND.(ERROR.LT.TOL)) FINISHED=.TRUE.
C
C          ***********************************************************
C          ** CHECK TO SEE IF WE HAVE REACHED THE MAXIMUM NUMBER **
C          ** OF SUBSECTIONS ALLOWED BY USER                     **
C          ***********************************************************
C
           IF(I.EQ.NMAX) FINISHED=.TRUE.
C
           I=I+1
C
           OLD=NEW
C
        ENDDO
C
        GR=DGDN*R(ZPRM,AE,BE)
C
        RETURN
        END
CC
C          ***************************************
C          ** IMAGINARY PART OF GREEN'S FUNCTION **
C          ***************************************
C
        REAL FUNCTION GI(Z,ZPRM,AE,BE,H)
        IMPLICIT NONE
        REAL PI,K,Z,ZPRM,R,YO,AE,BE,H,RX,RY,RCX,RCY
        REAL ZF,ZB,TX,TY,NX,NY,HX,HY,RPHX,RPHY,RMHX,RMHY,MPLUS,MMINUS
        REAL FPLUS,FMINUS,DGDN,C1,CP(10,10),NEW,OLD,TOL,ERROR
        INTEGER I,NS,Q,NMAX
        LOGICAL FINISHED
C
        PI=ACOS(-1.0)
        K=2.0*PI
        FINISHED=.FALSE.
        TOL=1.0E-5
        NMAX=8
C
C
C          *********************************************
C          ** CALCULATE X AND Y COMPONENTS OF RPRIME **
C          *********************************************
```

```
      RX=R(ZPRM,AE,BE)*COS(ZPRM)
      RY=R(ZPRM,AE,BE)*SIN(ZPRM)
C
C
C     *********************************************************
C     ** CALCULATE X AND Y COMPONENTS OF RC (MATCH POINT ANGLE) **
C     *********************************************************
      RCX=R(Z,AE,BE)*COS(Z)
      RCY=R(Z,AE,BE)*SIN(Z)
C
C
C     ***************************************************
C     ** CALCULATE BOUNDING ANGLES OF PRESENT SEGMENT **
C     ** TO FIND APPROXIMATE TANGENT VECTOR           **
C     ***************************************************
C
      ZF=ZPRM+0.5*H
      ZB=ZPRM-0.5*H
C
C
C         *********************************************************
C         ** CALCULATE APPROXIMATE TANGENT VECTOR COMPONENTS **
C         *********************************************************
      TX=R(ZF,AE,BE)*COS(ZF)-R(ZB,AE,BE)*COS(ZB)
      TY=R(ZF,AE,BE)*SIN(ZF)-R(ZB,AE,BE)*SIN(ZB)
C
C
C         *********************************************************
C         ** CALCULATE NORMAL VECTOR WHICH IS T CROSSED WITH Z **
C         ** DIVIDED BY ITS MAGNITUDE TO CREATE UNIT VECTOR    **
C         *********************************************************
      NX=TY/SQRT(TX**2+TY**2)
      NY=-TX/SQRT(TX**2+TY**2)
C
C
C     *****************************************************************
C     ** APPROXIMATE DERIVATIVE OF GREEN'S FUNCTION OF SOURCE POINT **
C     ** WITH RESPECT TO NORMAL DIRECTION TO CONTOUR               **
C     *****************************************************************
C
      I=1
C
      DO WHILE(.NOT.FINISHED)
C
      NS=2**I
C
      HX=NX/FLOAT(NS)
      HY=NY/FLOAT(NS)
C
C
C     ********************************************
C     ** CALCULATE RPRIME + SCALED NORMAL VECTOR **
C     ********************************************
C
      RPHX = RX+HX
      RPHY = RY+HY
C
C
C     ********************************************
C     ** CALCULATE RPRIME - SCALED NORMAL VECTOR **
C     ********************************************
      RMHX = RX-HX
```

```
      RMHY = RY-HY
C
C
C     ****************************************
C     ** CALCULATE APPROXIMATE DERIVATIVE **
C     ****************************************
      MPLUS=SQRT((RCX-RMHX)**2+(RCY-RMHY)**2)
      MMINUS=SQRT((RCX-RPHX)**2+(RCY-RPHY)**2)
C
      FPLUS=YO(K*(MPLUS))
      FMINUS=YO(K*(MMINUS))
C
      DGDN=(FPLUS-FMINUS)/(2.0*SQRT(HX*HX+HY*HY))
C
C
C
C
C     ****************************************************
C     ** ASSIGN SUMATION TO ARRAY FOR EXTRAPOLATION **
C     ****************************************************
      CP(1,I)=DGDN
C
C
C     *******************************************
C     ** RICHARDSON EXTRAPOLATION CALCULATION **
C     *******************************************
C
      C1=1.0
C
      DO Q=2,I
        C1=2*C1
        CP(Q,I)=(C1*CP(Q-1,I)-CP(Q-1,I-1))/(C1-1.0)
      ENDDO
C
      DGDN=CP(I,I)
C
      NEW=CP(I,I)
C
C
C     *************************************
C     ** CALCULATE ERROR AFTER FIRST PASS **
C     *************************************
C
      IF((I.GT.1).AND.(ABS(NEW).GT.0.0)) ERROR=ABS(OLD-NEW)/ABS(NEW)
C
C     ***********************************************************
C     ** CHECK ERROR AFTER FIRST PASS TO SEE IF INTEGRATION  **
C     ** IS WITHIN TOLERANCE END CALCULATION IF IT IS        **
C     ***********************************************************
C
      IF((I.GT.1).AND.(ERROR.LT.TOL)) FINISHED=.TRUE.
C
C
C     ***********************************************************
C     ** CHECK TO SEE IF WE HAVE REACHED THE MAXIMUM NUMBER **
C     ** OF SUBSECTIONS ALLOWED BY USER                     **
C     ***********************************************************
C
      IF(I.EQ.NMAX) FINISHED=.TRUE.
```

```
C
         I=I+1
C
         OLD=NEW
C
      ENDDO
C
      GI=-DGDN*R(ZPRM,AE,BE)*R(ZPRM,AE,BE)
C
      RETURN
      END
CC
C     ****************************
C     ** INCIDENT H-FIELD VALUES **
C     ****************************
C
      COMPLEX FUNCTION HINC(X,AE,BE,INC)
C
      IMPLICIT NONE
C
      REAL X,AE,BE,INC,PI,K,MAG,TOP,R
C
      COMPLEX JAY
C
      JAY=(0.0,1.0)
C
      PI=ACOS(-1.0)
      K=2.0*PI
C
      MAG=R(X,AE,BE)
C
      TOP=MAG*COS(X)*COS(INC)+MAG*SIN(X)*SIN(INC)
C
      HINC=COS(K*TOP)+JAY*SIN(K*TOP)
C
C     ********************************************
C     ** RELATE TO 1 VOLT/METER INCIDENT FIELD **
C     ********************************************
C
      HINC=HINC/376.73
C
      RETURN
      END
CC
C     ********************************************************
C     ** MAGNITUDE OF RADIUS VECTOR AS A FUNCTION OF THETA **
C     ********************************************************
C
      REAL FUNCTION R(THETA,AE,BE)
C
      IMPLICIT NONE
C
      REAL THETA,MAG,AE,BE
C
```

```
      MAG=1.0/( (COS(THETA)/BE)**2 + (SIN(THETA)/AE)**2 )
C
      R=SQRT(MAG*(COS(THETA)**2+SIN(THETA)**2))
C
      RETURN
      END
C
C
C     **********************************************************
C     ** MAGNITUDE OF THE DIFFERENCE OF TWO RADIUS VECTORS **
C     **********************************************************
C
      REAL FUNCTION DELTR(THETA1,THETA2,AE,BE)
C
      IMPLICIT NONE
C
      REAL THETA1,THETA2,MAG1,MAG2,IHAT,JHAT,AE,BE,R
C
      MAG1=R(THETA1,AE,BE)
C
      MAG2=R(THETA2,AE,BE)
C
      IHAT=MAG1*COS(THETA1)-MAG2*COS(THETA2)
C
      JHAT=MAG1*SIN(THETA1)-MAG2*SIN(THETA2)
C
      DELTR=SQRT(IHAT**2+JHAT**2)
C
      RETURN
      END
C
C
C     ****************************************************************
C     ** THIS SUBROUTINE APPROXIMATES THE INTEGRAL OF A FUNCTION  **
C     ** USING MIDPOINT INTEGRATION WITH RICHARDSON EXTRAPOLATION **
C     ****************************************************************
C
      REAL FUNCTION MIDINT(F,A,B,ZM,AE,BE,H)
C
      IMPLICIT NONE
      REAL CP(10,10),C1,SUMX,F,AE,BE
      REAL NSX,DX,A,B,ZM,MIDPNT,ERROR,OLD,NEW,TOL,H
      INTEGER COUNTX,I,Q,NMAX
      LOGICAL FINISHED
      EXTERNAL F
C
      FINISHED=.FALSE.
      COUNTX=1
      TOL=1.0E-4
      NMAX=8
C
      DO WHILE(.NOT.FINISHED)
C
C        ***********************************
C        ** CALCULATE NUMBER OF SUBSECTIONS **
C        ***********************************
         NSX=2.0**COUNTX
```

```
C
C        ****************************
C        ** CALCULATE STEP-SIZE **
C        ****************************
         DX=(B-A)/NSX

C        ******************************************************
C        ** EVALUATE AND SUM FUNCTION VALUE AT MIDPOINTS **
C        ******************************************************

         SUMX=0.0

         DO I=1,NSX
           MIDPNT=A+FLOAT(I-1)*DX+DX/2.0
           SUMX=SUMX+F(ZM,MIDPNT,AE,BE,H)
         ENDDO

C
C        **************************************************
C        ** ASSIGN SUMATION TO ARRAY FOR EXTRAPOLATION **
C        **************************************************
         CP(1,COUNTX)=SUMX*DX

C        ****************************************
C        ** RICHARDSON EXTRAPOLATION CALCULATION **
C        ****************************************

         C1=1.0

         DO Q=2,COUNTX
           C1=2*C1
           CP(Q,COUNTX)=(C1*CP(Q-1,COUNTX)-CP(Q-1,COUNTX-1))/(C1-1.0)
         ENDDO

         MIDINT=CP(COUNTX,COUNTX)

         NEW=CP(COUNTX,COUNTX)

C        ****************************************
C        ** CALCULATE ERROR AFTER FIRST PASS **
C        ****************************************

         IF(COUNTX.GT.1) ERROR=ABS(OLD-NEW)/ABS(NEW)

C        ************************************************************
C        ** CHECK ERROR AFTER FIRST PASS TO SEE IF INTEGRATION **
C        ** IS WITHIN TOLERANCE END CALCULATION IF IT IS        **
C        ************************************************************

         IF((COUNTX.GT.1).AND.(ERROR.LT.TOL)) FINISHED=.TRUE.

C        *********************************************************
C        ** CHECK TO SEE IF WE HAVE REACHED THE MAXIMUM NUMBER **
C        ** OF SUBSECTIONS ALLOWED BY USER                     **
C        *********************************************************
```

```
C
        IF(COUNTX.EQ.NMAX) FINISHED=.TRUE.
C
        COUNTX=COUNTX+1
C
        OLD=NEW
C

C
      END DO
C
      RETURN
      END
C
C
C       **********************************
C       ** BESSEL FUNCTION SUBROUTINE JO **
C       **********************************
      FUNCTION JO(X)
      IMPLICIT NONE
      INTEGER I
      REAL AK(7),BK(7),AA,BB,A,X,JO,PRODUCT
C
      DATA AK/0.79788456,-0.00000077,-0.00552740,
     ;       -0.00009512,0.00137237,-0.00072805,
     ;        0.00014476/
C
      DATA BK/-0.78539816,-0.04166397,-0.00003954,
     ;        0.00262573,-0.00054125,-0.00029333,
     ;        0.00013558/
C
      IF(ABS(X).LE.3.0)THEN
        A=(X/2.0)*(X/2.0)
        PRODUCT=1.0-A/49.0

        DO I=6,1,-1
          PRODUCT=1.0-A/FLOAT(I*I)*PRODUCT
        ENDDO

        JO=PRODUCT

      ELSE
C
        A=3.0/X
        AA=A*AK(7)
        BB=A*BK(7)
C
        DO I=6,1,-1
          AA=AK(I)+A*AA
          BB=BK(I)+A*BB
        ENDDO
C
        BB=BB+X
        JO=AA/SQRT(X)*COS(BB)
```

```
      ENDIF

      RETURN
      END
C     **********************************
C     ** BESSEL FUNCTION SUBROUTINE YO **
C     **********************************
      FUNCTION YO(X)
      IMPLICIT NONE
      INTEGER I
      REAL AT(7),AK(7),BK(7),BO,JO,GAMMA,PI,A,X,PRODUCT,AA,BB,YO
C
      DATA AK/0.79788456,-0.00000077,-0.00552740,
     ;      -0.00009512,0.00137237,-0.00072805,
     ;       0.00014476/
C
      DATA BK/-0.78539816,-0.04166397,-0.00003954,
     ;       0.00262573,-0.00054125,-0.00029333,
     ;       0.00013558/
C
      GAMMA=0.57721567
      PI=ACOS(-1.0)
      A=(X/2.0)*(X/2.0)

      IF(ABS(X).LE.3)THEN
        AT(1)=1.0

        DO I=2,7
          AT(I)=AT(I-1)+1.0/FLOAT(I)
        ENDDO

        BO=2.0*(GAMMA+LOG(X/2.0))*JO(X)

        PRODUCT=AT(6)-A*AT(7)/49.0

        DO I=6,2,-1
          PRODUCT=AT(I-1)-A/FLOAT(I*I)*PRODUCT
        ENDDO

        YO=(BO+ 2.0*A*PRODUCT)/PI

      ELSE
C
        A=3.0/X
        AA=A*AK(7)
        BB=A*BK(7)
C
        DO I=6,1,-1
          AA=AK(I)+A*AA
          BB=BK(I)+A*BB
        ENDDO
C
        BB=BB+X
        YO=AA/SQRT(X)*SIN(BB)
```

```
      ENDIF
C
      RETURN
      END
```

# Appendix E:
# Chapter 6
# FORTRAN Computer Programs

```
C    **************************************************************
C    ** THIS PROGRAM CALCULATES THE RADAR CROSS SECTION OF  **
C    ** A THIN METAL PLATE USING THE MOMENT METHOD.         **
C    ** PULSE EXPANSION FUNCTIONS ARE USED FOR JX, JY AND   **
C    ** THE CHARGE. THE FIELD IS MATCHED FOR EX AND EY AT   **
C    ** THE CENTER OF EACH CURRENT PULSE.                   **
C    **                                                     **
C    ** THIS PROGRAM IMPLEMENTS (6.34) AND (6.35)           **
C    **                                                     **
C    ** RANDY BANCROFT  8/19/95                             **
C    **                                                     **
C    **************************************************************
C    **                                                     **
C    ** VARIABLE DICTIONARY (MATRIX SECTION NOT INCLUDED)   **
C    **                (INTEGRATION SECTIONS NOT INCLUDED)  **
C    **                                                     **
C    ** A -- MOMENT MATRIX                                  **
C    ** A1 -- RETURN VARIABLE FOR MAGNETIC VECTOR POTENTIAL **
C    **        TERM INTEGRATIONS                            **
C    ** B -- ENFORCEMENT MATRIX (BECOMES SOLUTION MATRIX)   **
C    ** C -- SPEED OF LIGHT IN FREE SPACE                   **
C    ** C1 -- CONSTANT USED IN RICHARDSON'S EXTRAPOLATION   **
C    ** CP -- ARRAY FOR RICHARDSON'S EXTRAPOLATION          **
C    ** CSUM -- VARIABLE FOR COMPLEX SUMMATION              **
C    ** EO -- PERMITIVITTY OF FREE SPACE                    **
C    ** ELEMENT -- TEMPORARY VARIABLE FOR A(I,J)            **
C    ** EPHI -- ELECTRIC FIELD OF PHI COMPONENT OF          **
C    **            INCOMING ELECTROMAGNETIC WAVE (COMPLEX)  **
C    ** ETA -- CHARACTERISTIC IMPEDANCE OF FREE SPACE       **
C    ** ETHETA -- ELECTRIC FIELD OF THETA COMPONENT OF      **
C    **            INCOMING ELECTROMAGNETIC WAVE (COMPLEX)  **
C    ** EXINC -- ELECTRIC FIELD ALONG X AT MATCH POINT      **
C    ** EYINC -- ELECTRIC FIELD ALONG Y AT MATCH POINT      **
C    ** FREQ -- FREQUENCY OF PLANE WAVE                     **
C    ** H -- SEGMENT LENGTH                                 **
C    ** I -- LOOP VARIABLE                                  **
C    ** IJ -- MATRIX ARGUMENT TO KEEP TRACK OF JX AND JY    **
C    ** INC -- INCIDENT ANGLE OF PLANE WAVE                 **
C    ** IX -- X INTEGRATION FOR EX FIELD USED FOR RCS CALC  **
C    ** IY -- Y INTEGRATION FOR EY FIELD USED FOR RCS CALC  **
C    ** J -- LOOP VARIABLE                                  **
C    ** JAY -- SQUARE ROOT OF -1                            **
C    ** JX -- ARRAY TO HOLD X DIRECTED CURRENTS             **
C    ** JY -- ARRAY TO HOLD Y DIRECTED CURRENTS             **
C    ** K -- FREE SPACE WAVENUMBER                          **
C    ** L -- LENGTH OF SQUARE PLATE SIDES                   **
C    ** LAMBDA -- FREE SPACE WAVELENGTH                     **
C    ** M -- LOOP VARIABLE                                  **
C    ** MN -- MATRIX ARGUMENT TO KEEP TRACK OF JX AND JY    **
C    ** MTX -- ARRAY SIZE [A],[B]                           **
C    ** N -- LOOP VARIABLE                                  **
C    ** NMAX -- MAXIMUM NUMBER OF SEGMENT DIVISIONS         **
C    ** NS -- NUMBER OF SEGMENTS                            **
C    ** OMEGA -- ANGULAR FREQUENCY OF PLANE WAVE            **
```

```
C     ** PHASE -- PHASE OF EX OR EY FIELD AT MATCH POINT     **
C     ** PHI -- INCIDENT ANGLE OF INCOMING PLANE WAVE         **
C     ** PI -- RATIO OF CIRCUMFERENCE TO DIAMETER OF CIRCLE   **
C     ** Q -- RICHARDSON'S EXTRAPOLATION LOOP VARIABLE        **
C     ** S -- SEGMENT LOOP COUNTER                            **
C     ** S1,S2,S3,S4 -- RETURN VARIABLES FOR INTEGRATION TO   **
C     **                PRODUCE SCALAR POTENTIAL TERMS        **
C     ** SIGMA1 -- THETA NORMALIZED RADAR CROSS SECTION       **
C     ** SIGMA2 -- PHI NORMALIZED RADAR CROSS SECTION         **
C     ** T1 -- TEMPORARY VARIABLE                             **
C     ** THETA -- INCIDENT ANGLE OF INCOMING PLANE WAVE       **
C     ** UO -- PERMEABILITY OF FREE SPACE                     **
C     ** XM,XMP1,XMM1,XMMHALF,XMPHALF -- INTEGRATION LIMITS   **
C     **                              ALONG X OF PULSE        **
C     **                                                      **
C     ** XMTCH -- X COORDINATE OF MATCH POINT                 **
C     ** XMTCHM1 -- X COORDINATE OF MATCH POINT MINUS ONE     **
C     **              SEGMENT LENGTH                          **
C     ** XMTCHMH -- X COORDINATE OF MATCH PONT MINUS HALF     **
C     **              OF A SEGMENT LENGTH                     **
C     ** XMTCHP1 -- X COORDINATE OF MATCH POINT PLUS ONE      **
C     **              SEGMENT LENGTH                          **
C     ** XMTCHPH -- X COORDIANTE OF MATCH POINT PLUSE HALF    **
C     **              OF A SEGMENT LENGTH                     **
C     ** Y -- ARRAY WHICH HOLDS NUMBER OF SEGMENTS FOR S      **
C     ** YMTCH -- Y COORDINATE OF MATCH POINT                 **
C     ** YMTCHM1 -- Y COORDINATE OF MATCH POINT MINUS ONE     **
C     **              SEGMENT LENGTH                          **
C     ** YMTCHMH -- Y COORDINATE OF MATCH POINT MINUS HALF    **
C     **              OF A SEGMENT LENGTH                     **
C     ** YMTCHP1 -- Y COORDINATE OF MATCH POINT PLUS ONE      **
C     **              SEGMENT LENGTH                          **
C     ** YMTCHPH -- Y COORDINATE OF MATCH POINT PLUS HALF     **
C     **              OF A SEGMENT LENGTH                     **
C     ** YN,YNP1,YNM1,YNMHALF,YNPHALF -- INTEGRATION LIMITS   **
C     **                              ALONG Y OF PULSE        **
C     ***********************************************************
C
      PROGRAM PLATE
C
      IMPLICIT NONE
C
      INTEGER MTX
      PARAMETER(MTX=144)
C
C     ***************************
C     ** DECLARE REAL VARIABLES **
C     ***************************
C
      REAL PI,EO,UO,C,LAMBDA,K,L,E,SIGMA1,SIGMA2,FREQ,OMEGA
      REAL CP(10,10),CP1(10,10),C1,THETA,PHI,XMTCH,YMTCH,XMTCHPH,XMP1
      REAL XMTCHMH,YMTCHPH,YMTCHMH,XM,XMPHALF,XMMHALF,YN,YNP1
      REAL YNM1,YNMHALF,YNPHALF,XMM1,MAX
C
C     ***************************
```

```
C       ** DECLARE COMPLEX VARIBLES **
C       ******************************
C
        COMPLEX JAY,A(MTX,MTX),B(MTX),ETHETA,EPHI,EXINC,EYINC,PHASE,T1,IX
        COMPLEX IY,S1,S2,S3,S4,A1,JX(MTX,MTX),JY(MTX,MTX),ATEMP,BTEMP
        COMPLEX X(MTX),SUM,MJK
C
C       ******************************
C       ** DECLARE INTEGER VARIBLES **
C       ******************************
C
        INTEGER K2,ICOL,IROW,COL,XTMP,XINDX(MTX),I,J,N,M,S,NMAX,NS,IJ,MN,Q
C
C       *********************************
C       ** ASSIGN FUNDAMENTAL CONSTANTS **
C       *********************************
C
        JAY=(0.0,1.0)
        PI=ACOS(-1.0)
        C=2.997925E8
        EO=8.854223E-12
        UO=1.256640E-6
C
C       ****************************
C       ** INPUT DESIRED OPTIONS **
C       ****************************
C
C
        WRITE (*,'(5X,A)')'INPUT LENGTH OF PLATE SIDE (WAVELENGTHS) '
        READ (*,*) L
C
        WRITE (*,'(5X,A)')'INPUT INCIDENT ANGLE OF PLANE WAVE (THETA) '
        READ (*,*) THETA
        THETA=THETA*PI/180.0
C
        WRITE (*,'(5X,A)')'INPUT INCIDENT ANGLE OF PLANE WAVE (PHI) '
        READ (*,*) PHI
        PHI=PHI*PI/180
C
C       *************************
C       ** USE 1 LAMBDA VALUES **
C       *************************
C
        FREQ=C
        OMEGA=2.0*PI*FREQ
        LAMBDA=C/FREQ
        K=2.0*PI/LAMBDA
C
C       **********************************************
C       ** SET ELECTRIC FIELD VALUES ETHETA AND EPHI **
C       **********************************************
        ETHETA=(1.0,0.0)
        EPHI=(0.0,0.0)
C
C       **********************************************
```

```
C      ** MAXIMUM VALUE OF S WHICH DETERMINES    **
C      ** THE MAXIMUM NUMBER OF CURRENT SEGMENTS **
C      ********************************************
C
       NMAX=3
C
C      *********************************
C      ** OUTER LOOP CONTROLS HOW MANY **
C      ** SEGMENT DIVISIONS OCCUR      **
C      *********************************
C
       DO S=1,NMAX
C
C        ***************************
C        ** CALCULATE THE NUMBER  **
C        ** OF SEGMENTS FOR THIS  **
C        ** LOOP ITERATION        **
C        ***************************
C
         NS=2**S
C
C        ***************************
C        ** CALCULATE STEP LENGTH **
C        ***************************
C
       H=L/FLOAT(NS+1)
C
C        *********************************
C        ** FILL THE [A] AND [B] MATRIX **
C        *********************************
C
C      ************************************************************
C      ** EQUATION SET ONE (6-32) PRODUCES THE TOP HALF **
C      ** OF THE MOMENT MATRIX BY ENFORCING THE VALUE   **
C      ** OF Ex AT EACH MATCH POINT FOR Jx AND Jy       **
C      ************************************************************
C
       MN=0
       IJ=0
C
C      *******************************************
C      ** THIS LOOP SET CALCULATES THE MATCH    **
C      ** POINT FOR THE TOP HALF OF THE MATRIX  **
C      *******************************************
C
       DO J=1,NS+1
         DO I=1,NS
C
C          *****************************
C          ** CALCULATE MATCH POINT FOR **
C          ** CURRENT AT POINT I,J     **
C          *****************************
C
         XMTCH=FLOAT(I)*H
         XMTCHPH=XMTCH+H/2.0
```

```
      XMTCHMH=XMTCH-H/2.0
C
      YMTCH=FLOAT(J)*H
      YMTCHPH=YMTCH+H/2.0
      YMTCHMH=YMTCH-H/2.0
C
C
C     *****************************************
C     ** CALCULATE THE INTEGRATION LIMITS   **
C     ** FOR THE INTEGRATION OF THE UNKNOWN **
C     ** CURRENT AND CHARGE                 **
C     *****************************************
C
      IJ=IJ+1
C
C
C     *****************************************
C     ** X COMPONENT OF CURRENT AT POINT M,N **
C     *****************************************
C
      MN=0
C
      DO N=1,NS+1
       DO M=1,NS
C
        MN=MN+1
C
C       *************************
C       ** INTEGRATION LIMITS **
C       *************************
C
        XM=FLOAT(M)*H
        XMP1=FLOAT(M+1)*H
        XMM1=FLOAT(M-1)*H
        XMMHALF=XM-H/2.0
        XMPHALF=XM+H/2.0
C
        YN=FLOAT(N)*H
        YNP1=FLOAT(N+1)*H
        YNM1=FLOAT(N-1)*H
        YNMHALF=YN-H/2.0
        YNPHALF=YN+H/2.0
C
C       ***********************************************
C       ** VECTOR POTENTIAL TERM OF Z-MATRIX FOR JX **
C       ***********************************************
C
        CALL INTEGRAL(XMMHALF,XMPHALF,YNM1,YN,XMTCH,YMTCHMH,A1)
C
C       ***********************************************
C       ** SCALAR POTENTIAL TERMS OF Z-MATRIX FOR JX **
C       ***********************************************
C
C       ****************************
C       ** FIRST TERM INTEGRATION **
C       ****************************
```

```
      CALL INTEGRAL(XMM1,XM,YNM1,YN,XMTCHPH,YMTCHMH,S1)
C
C
C     ****************************
C     ** SECOND TERM INTEGRATION **
C     ****************************
C
      CALL INTEGRAL(XM,XMP1,YNM1,YN,XMTCHPH,YMTCHMH,S2)
C
C     ****************************
C     ** THIRD TERM INTEGRATION **
C     ****************************
C
      CALL INTEGRAL(XMM1,XM,YNM1,YN,XMTCHMH,YMTCHMH,S3)
C
C     ****************************
C     ** FOURTH TERM INTEGRATION **
C     ****************************
C
      CALL INTEGRAL(XM,XMP1,YNM1,YN,XMTCHMH,YMTCHMH,S4)
C
C
C     **************************************
C     ** X DIRECTED CURRENT MATRIX ELEMENT **
C     **************************************
C
      A(IJ,MN)=(-S1+S2+S3-S4)*JAY/(4.0*PI*EO*OMEGA*H)
    ;                     -JAY*OMEGA*UO*A1/(4.0*PI)*H
C
C     **************************************
C     ** PRINT THE CURRENT [A] MATRIX VALUE **
C     **************************************
C
      WRITE (*,'(A1,4X,A,I3,I3,A,F15.4,2X,F15.4)')'+','Ax(',IJ,MN,
    ;  ')=',A(IJ,MN)
C
C
      ENDDO
      ENDDO        !MN LOOP EQUATION 1 (6-32) JX
C
C
C     ***********************************************************
C     ** CALCULATE THE MATRIX VALUES FOR THE Jy CURRENTS OF **
C     ** EQUATION 1                                          **
C     ***********************************************************
C
      DO N=1,NS
       DO M=1,NS+1
C
       MN=MN+1
C
C     **********************
C     ** INTEGRATION LIMITS **
C     **********************
      XM=FLOAT(M)*H
      XMP1=FLOAT(M+1)*H
      XMM1=FLOAT(M-1)*H
```

```
           XMMHALF=XM-H/2.0
           XMPHALF=XM+H/2.0
C
           YN=FLOAT(N)*H
           YNP1=FLOAT(N+1)*H
           YNM1=FLOAT(N-1)*H
           YNMHALF=YN-H/2.0
           YNPHALF=YN+H/2.0
C
C
C       ****************************************************
C       ** SCALAR POTENTIAL TERM OF Z-MATRIX FOR JY     **
C       ** FIRST EQUATION HAS NO VECTOR POTENTIAL TERM **
C       ****************************************************
C
C
C       ****************************
C       ** SCALAR POTENTIAL TERMS JY **
C       ****************************
C
C
C       ****************************
C       ** FIRST TERM INTEGRATION **
C       ****************************
C
           CALL INTEGRAL(XMM1,XM,YNM1,YN,XMTCHPH,YMTCHMH,S1)
C
C       ****************************
C       ** SECOND TERM INTEGRATION **
C       ****************************
C
           CALL INTEGRAL(XMM1,XM,YN,YNP1,XMTCHPH,YMTCHMH,S2)
C
C       ****************************
C       ** THIRD TERM INTEGRATION **
C       ****************************
C
           CALL INTEGRAL(XMM1,XM,YNM1,YN,XMTCHMH,YMTCHMH,S3)
C
C       ****************************
C       ** FOURTH TERM INTEGRATION **
C       ****************************
C
           CALL INTEGRAL(XMM1,XM,YN,YNP1,XMTCHMH,YMTCHMH,S4)
C
           A(IJ,MN)=(-S1+S2+S3-S4)*JAY/(4.0*PI*EO*OMEGA*H)
C
C       ****************************************
C       ** PRINT THE CURRENT [A] MATRIX VALUE **
C       ****************************************
C
           WRITE (*,'(A1,4X,A,I3,I3,A,F15.4,2X,F15.4)')')'+','Ay(',IJ,MN,
      ;')=',A(IJ,MN)
C
C
           ENDDO ! N LOOP
           ENDDO ! M LOOP ** END OF EQUATION SET ONE (6-32) Jy **
```

```
C
C
C          ***********************************
C          ** FILL THE FIRST HALF THE [B] MATRIX **
C          ***********************************
C
C          ***********************************
C          ** CALCULATE FIELD AT MATCH POINT **
C          ***********************************
C
           EXINC=ETHETA*COS(THETA)*COS(PHI)-EPHI*SIN(PHI)
           EYINC=ETHETA*COS(THETA)*SIN(PHI)+EPHI*COS(PHI)
C
           PHASE=CEXP(JAY*K*(XMTCH*SIN(THETA)*COS(PHI)
     ;                 +YMTCHMH*SIN(THETA)*SIN(PHI)))
C
           EXINC=EXINC*PHASE
           EYINC=EYINC*PHASE
C
           B(IJ)=-EXINC*H
C
      ENDDO ! J LOOP
      ENDDO ! I LOOP    ** END OF EQUATION SET 1 (6-32)  **
C
C
C
C
C      ***********************************
C      ** SECOND EQUATION SET (6-33) IS  **
C      ** NEEDED TO OBTAIN SQUARE MATRIX **
C      ***********************************
C
C      ***********************************
C      ** THIS OUTER LOOP SET CALCULATES **
C      ** THE MATCH POINT FOR THE TOP    **
C      ** HALF OF THE MATRIX             **
C      ***********************************
       DO J=1,NS
       DO I=1,NS+1
C
C      ***********************************
C      ** CALCULATE MATCH POINT FOR CURRENT **
C      ** AT POINT I,J                       **
C      ***********************************
C
       XMTCH=FLOAT(I)*H
       XMTCHPH=XMTCH+H/2.0
       XMTCHMH=XMTCH-H/2.0
C
       YMTCH=FLOAT(J)*H
       YMTCHPH=YMTCH+H/2.0
       YMTCHMH=YMTCH-H/2.0
C
C      ***********************************
C      ** CALCULATE THE INTEGRATION LIMITS  **
C      ** FOR THE INTEGRATION OF THE UNKNOWN **
```

```
C       ** CURRENT AND CHARGE                    **
C       *****************************************
C
        IJ=IJ+1
C
C       ****************************************************
C       ** X COMPONENT OF CURRENT AT POINT M,N (6-33) **
C       ****************************************************
C
        MN=0
C
        DO N=1,NS+1
         DO M=1,NS
C
          MN=MN+1
C
C         ***********************
C         ** INTEGRATION LIMITS **
C         ***********************
          XM=FLOAT(M)*H
          XMP1=FLOAT(M+1)*H
          XMM1=FLOAT(M-1)*H
          XMMHALF=XM-H/2.0
          XMPHALF=XM+H/2.0
C
          YN=FLOAT(N)*H
          YNP1=FLOAT(N+1)*H
          YNM1=FLOAT(N-1)*H
          YNMHALF=YN-H/2.0
          YNPHALF=YN+H/2.0
C
C         **************************************************************
C         ** SCALAR POTENTIAL TERMS OF A-MATRIX FOR JX (6-33)   **
C         ** SECOND EQUATION SET HAS NO VECTOR POTENTIAL FOR JX **
C         **************************************************************
C
C         ***************************
C         ** FIRST TERM INTEGRATION **
C         ***************************
          CALL INTEGRAL(XMM1,XM,YNM1,YN,XMTCHMH,YMTCHPH,S1)
C
C         ***************************
C         ** SECOND TERM INTEGRATION **
C         ***************************
C
          CALL INTEGRAL(XM,XMP1,YNM1,YN,XMTCHMH,YMTCHPH,S2)
C
C         ***************************
C         ** THIRD TERM INTEGRATION **
C         ***************************
C
          CALL INTEGRAL(XMM1,XM,YNM1,YN,XMTCHMH,YMTCHMH,S3)
C
C         ***************************
```

```
C        ** FOURTH TERM INTEGRATION **
C        *****************************
C
         CALL INTEGRAL(XM,XMP1,YNM1,YN,XMTCHMH,YMTCHMH,S4)
C
C        *****************************************
C        ** X DIRECTED CURRENT MATRIX ELEMENT **
C        *****************************************
C
         A(IJ,MN)=(-S1+S2+S3-S4)*JAY/(4.0*PI*EO*OMEGA*H)
C
C        *****************************************
C        ** PRINT THE CURRENT [A] MATRIX VALUE **
C        *****************************************
C
         WRITE (*,'(A1,4X,A,I3,I3,A,F15.4,2X,F15.4)')'+','Ax(',IJ,MN,
     ;        ')=',A(IJ,MN)
C
C
      ENDDO
      ENDDO          !MN LOOP EQUATION 2 (6-33) JX
C
C     **************************************************************
C     ** CALCULATE THE Jy MATRIX ELEMENTS FOR EQUATION SET 2 **
C     **************************************************************
C
      DO N=1,NS
       DO M=1,NS+1
C
         MN=MN+1
C
C        **************************
C        ** INTEGRATION LIMITS **
C        **************************
         XM=FLOAT(M)*H
         XMP1=FLOAT(M+1)*H
         XMM1=FLOAT(M-1)*H
         XMMHALF=XM-H/2.0
         XMPHALF=XM+H/2.0
C
C
         YN=FLOAT(N)*H
         YNP1=FLOAT(N+1)*H
         YNM1=FLOAT(N-1)*H
         YNMHALF=YN-H/2.0
         YNPHALF=YN+H/2.0
C
C
C
C
C        ***********************************************
C        ** VECTOR POTENTIAL TERM OF Z-MATRIX FOR JY **
C        ***********************************************
C
         CALL INTEGRAL(XMM1,XM,YNMHALF,YNPHALF,XMTCHMH,YMTCH,A1)
C
```

```
C
C
C          ******************************
C          ** SCALAR POTENTIAL TERMS JY **
C          ******************************
C
C
C          ****************************
C          ** FIRST TERM INTEGRATION **
C          ****************************
C
           CALL INTEGRAL(XMM1,XM,YNM1,YN,XMTCHMH,YMTCHPH,S1)
C
C          *****************************
C          ** SECOND TERM INTEGRATION **
C          *****************************
C
           CALL INTEGRAL(XMM1,XM,YN,YNP1,XMTCHMH,YMTCHPH,S2)
C
C          ****************************
C          ** THIRD TERM INTEGRATION **
C          ****************************
C
           CALL INTEGRAL(XMM1,XM,YNM1,YN,XMTCHMH,YMTCHMH,S3)
C
C          *****************************
C          ** FOURTH TERM INTEGRATION **
C          *****************************
C
           CALL INTEGRAL(XMM1,XM,YN,YNP1,XMTCHMH,YMTCHMH,S4)
C
C
           A(IJ,MN)=(-S1+S2+S3-S4)*JAY/(4.0*PI*EO*OMEGA*H)
     ;            -JAY*OMEGA*UO*H*A1/(4.0*PI)
C
C          ******************************************
C          ** PRINT THE CURRENT [A] MATRIX VALUE **
C          ******************************************
C
           WRITE (*,'(A1,4X,A,I3,I3,A,F15.4,2X,F15.4)')')'+','Ay(',IJ,MN,
     ;')=',A(IJ,MN)
C
           ENDDO ! N LOOP
           ENDDO  ! M LOOP ** END OF EQUATION SET 2 (6-33) Jy **
C
C
C          ******************************************
C          ** FILL THE FIRST HALF THE [B] MATRIX **
C          ******************************************
C
C          ***********************************
C          ** CALCULATE FIELD AT MATCH POINT **
C          ***********************************
C
           EXINC=ETHETA*COS(THETA)*COS(PHI)-EPHI*SIN(PHI)
           EYINC=ETHETA*COS(THETA)*SIN(PHI)+EPHI*COS(PHI)
```

```
C

        PHASE=CEXP(JAY*K*(XMTCHMH*SIN(THETA)*COS(PHI)
;                     +YMTCH*SIN(THETA)*SIN(PHI)))
C

        EXINC=EXINC*PHASE
        EYINC=EYINC*PHASE
C
C

        B(IJ)=-EYINC*H

C

     ENDDO ! J LOOP   ** EQUATION SET 2 Ey **
     ENDDO ! I LOOP   **                    **


C
C
C     **********************************************
C     ** SET DUMMY VARIBLES FOR MATRIX ROUTINE **
C     **********************************************
C

      N=2*(NS)*(NS+1)
      M=2*(NS)*(NS+1)
C
C     ********************************
C     ** SOLVE FOR PLATE CURRENTS **
C     ********************************
C
C
C   ************************************************************
C   ** GAUSSIAN ELIMINATION ALGORITHM WITH TOTAL PIVOTING **
C   ************************************************************
C
C     ****************************************
C     ** INITIALIZE SOLUTION INDEX MATRIX **
C     ****************************************
C

      DO I=1,N
        XINDX(I)=I
      END DO
C
C
C     ****************************************
C     ** OUTER LOOP CONTROLS ELIMINATION **
C     ****************************************
C

      DO K2=1,N-1
C
C
C     ****************************************
C     ** SEARCH FOR MAXIMUM VALUE IN ARRAY **
C     ****************************************
C

      MAX=0.0
C

      DO I=K2,N
        DO J=K2,N
```

```
        IF(CABS(A(I,J)).GT.MAX)THEN
          MAX=A(I,J)
C         ************************************
C         ** KEEP INDICES OF MAXIMUM ELEMENT **
C         ************************************
          IROW=I
          ICOL=J
        ENDIF
      END DO
    END DO
C
C     *******************************************************
C     ** DETERMINE IF A ROW EXCHANGE OR COLUMN EXCHANGE **
C     ** IS REQUIRED TO BRING ELEMENT TO PIVOT          **
C     *******************************************************
C
C
C     *******************************************************
C     ** IF THE COLUMN INDEX AND THE ROW INDEX MATCH **
C     ** A ROW EXCHANGE WILL BRING THE MAXIMUM        **
C     ** ELEMENT TO THE PIVOT POINT                   **
C     *******************************************************
      IF(IROW.NE.K2)THEN
C       *******************************************************
C       ** EXCHANGE IROW (I.E. MAX ELEMENT ROW) WITH ROW K2 **
C       *******************************************************
        DO I=1,N
C         *******************************
C         ** FIRST SWAP THE [A] MATRIX **
C         *******************************
          ATEMP=A(IROW,I)
          A(IROW,I)=A(K2,I)
          A(K2,I)=ATEMP
        END DO
C         *******************************
C         ** THEN SWAP THE [B] MATRIX **
C         *******************************
        BTEMP=B(IROW)
        B(IROW)=B(K2)
        B(K2)=BTEMP
      ENDIF
C
C     ***********************************************************
C     ** EXCHANGE ICOL (I.E. MAX ELEMENT COLUMN WITH COLUMN K2 **
C     ***********************************************************
      IF(ICOL.NE.K2)THEN
C
        DO I=1,N
C         *******************************
C         ** FIRST SWAP THE [A] MATRIX **
C         *******************************
          ATEMP=A(I,ICOL)
          A(I,ICOL)=A(I,K2)
          A(I,K2)=ATEMP
        END DO
C         ***********************************
```

```
C            ** THEN SWAP THE [XINDX] MATRIX  **
C            ** WHICH DOES THE SOLUTION ORDER **
C            ** BOOK KEEPING                  **
C            ********************************************
             XTMP=XINDX(ICOL)
             XINDX(ICOL)=XINDX(K2)
             XINDX(K2)=XTMP
C
         ENDIF
C
C
C        ****************************************
C        ** NORMALIZE EACH ROW AND ELIMINATE **
C        ****************************************
C
         IF(CABS(A(K2,K2)).EQ.0.0) WRITE(*,*) 'ZERO PIVOT ENCOUNTERED: ERRO
        ;R 1'
C
         DO J=K2+1,N
           MJK=A(J,K2)/A(K2,K2)
           DO COL=K2,N
            A(J,COL)=A(J,COL)-MJK*A(K2,COL)
           END DO
           B(J)=B(J)-MJK*B(K2)
         END DO
C
         WRITE(*,'(A1,A,I3,A,I3,A,20X)')')'+','Matrix Row ',K2,' of ',N,
        ;' Total Rows'
C
         ENDDO
C
C        ************************
C        ** BACKSUBSTITUTION **
C        ************************
C
         X(N)=B(N)/A(N,N)
C
         DO I=(N-1),1,-1
           SUM=(0.0,0.0)
           DO J=(I+1),N
             SUM=SUM+A(I,J)*X(J)
           END DO
           X(I)=(B(I)-SUM)/A(I,I)
         END DO
C
C        ****************************************************************
C        ** USE THE INDEX ARRAY TO LOCATE PERMUTATED SOLUTIONS **
C        ****************************************************************
C
         DO I=1,N
           DO J=1,N
             IF(XINDX(J).EQ.I) B(I)=X(J)
           END DO
         END DO
C
```

```
C       *****************************
C       ** ERASE MATRIX PRINT LINE **
C       *****************************
        WRITE (*,'(A1,A)')'+',' 
     ;
C
C
C       ***********************************************
C       ** ARRANGE LINEAR ARRAY OF CURRENTS INTO TWO **
C       ** DIMENSIONAL ARRAY FOR OUTPUT AND RCS      **
C       ** CALCULATION                              **
C       ***********************************************
C
        MN=0
C
C       *******************************
C       ** X-DIRECTED CURRENT ON PLATE **
C       *******************************
        DO N =1,NS+1
         DO M= 1,NS
          MN=MN+1
          JX(M,N)=B(MN)
         ENDDO
        ENDDO
C
C       *******************************
C       ** Y-DIRECTED CURRENT ON PLATE **
C       *******************************
        DO N =1,NS
         DO M= 1,NS+1
          MN=MN+1
          JY(M,N)=B(MN)
         ENDDO
        ENDDO
C
C       **********************************************************
C       ** CALCULATE MAGNETIC VECTOR POTENTIAL INTEGRATION **
C       ** OF X AND Y COMPONENTS TO OBTAIN THE E-FIELD FOR **
C       ** LATER RCS CALCULATION                          **
C       **********************************************************
C
        IX=(0.0,0.0)
C
        DO M = 1,NS
         DO N =1,NS+1

         XM=M*H
         YNMHALF=N*H-H/2

         IX=IX+JX(M,N)*CEXP(JAY*K*(XM*SIN(THETA)*COS(PHI)
     ;                    +YNMHALF*SIN(THETA)*SIN(PHI)))
         ENDDO
        ENDDO

        IX=IX*H*H
```

```
      IY=(0.0,0.0)
C
      DO M = 1,NS+1
       DO N =1,NS

          XMMHALF=M*H-H/2.0
          YN=N*H

       IY=IY+JY(M,N)*CEXP(JAY*K*(XMMHALF*SIN(THETA)*COS(PHI)
      ;                    +YN*SIN(THETA)*SIN(PHI)))
       ENDDO
      ENDDO

      IY=IY*H*H
C
C     **********************************
C     ** CALCULATE RADAR CROSS SECTION **
C     **********************************
C
C     ***************
C     ** THETA RCS **
C     ***************
C
      T1=IX*COS(THETA)*COS(PHI)+IY*COS(THETA)*SIN(PHI)
      IF(CABS(ETHETA).NE.0.0)THEN
        SIGMA1=REAL(T1*CONJG(T1))/REAL(ETHETA*CONJG(ETHETA))
      ;                    *OMEGA*OMEGA*UO*UO/(4.0*PI)
      ELSE
        SIGMA1=0.0
      ENDIF
C     *************
C     ** PHI RCS **
C     *************
C
      T1=-IX*SIN(PHI)+IY*COS(PHI)
      IF(CABS(EPHI).NE.0.0)THEN
        SIGMA2=REAL(T1*CONJG(T1))/REAL(EPHI*CONJG(EPHI))
      ;                    *OMEGA*OMEGA*UO*UO/(4.0*PI)
      ELSE
        SIGMA2=0.0
      ENDIF
C
        CP1(1,S)=SIGMA1
C
        CP(1,S)=SIGMA2
C
C     ******************************************
C     ** RICHARDSON EXTRAPOLATION CALCULATION **
C     ******************************************
C
        C1=1
        DO Q=2,S
          C1=2*C1
```

```
         CP(Q,S)=(C1*CP(Q-1,S)-CP(Q-1,S-1))/(C1-1)
         ENDDO

         C1=1
         DO Q=2,S
           C1=2*C1
           CP1(Q,S)=(C1*CP1(Q-1,S)-CP1(Q-1,S-1))/(C1-1)
         ENDDO
C
C      *******************
C      ** PRINT RESULTS **
C      *******************
C
         IF(S.EQ.1)THEN

         DO I=1,25
           WRITE(*,*)' '
         ENDDO
C
         WRITE(*,'(5X,A,F5.3,A)')'RADAR CROSS SECTION OF A THIN METALLIC PL
        ;ATE ',L,' WAVELENGTHS SQUARE'
C
         WRITE(*,*)' '
C
         WRITE(*,'(5X,A,F6.2,18X,A,F6.2)')'INCIDENT ANGLE: Theta =',
        ;THETA*180.0/PI,'Phi =',PHI*180.0/PI
C
         WRITE(*,*)' '
C
         WRITE(*,'(5X,A,F6.2,5X,A,F6.2)')'INCIDENT FIELD COMPONENTS |Etheta
        ;| =',CABS(EPHI),'|Ephi| =',CABS(ETHETA)
C
         WRITE(*,*)' '
C
         WRITE(*,'(7X,A,4X,A,2X,A,8X,A)')'N','Sigma Theta','Sigma Phi',
        ;'Extrapolation'
C
         WRITE(*,*)' '
C
         ENDIF
C
         WRITE(*,'(5X,I3,7X,F5.2,7X,F5.2,10X,F5.2,2X,F5.2)')NS,
        ;SIGMA1,SIGMA2,CP1(S,S),CP(S,S)
C
         WRITE(*,*)'     '
C
         ENDDO ! SEGMENT LOOP
C
         END
CC
C      ***********************************************
C      ** THIS SUBROUTINE CALLS MIDPOINT INTEGRATION **
C      ** ROUTINES TO PERFORM A TWO DIMENSIONAL      **
C      ** INTEGRATION FOR THE REAL AND IMAGINARY     **
C      ** PARTS OF THE [A] MATRIX                    **
```

```
C     ** THE INTEGRATION IS FROM A TO B ALONG X    **
C     **                        C TO D ALONG Y    **
C     ** WITH MATCH POINT XMATCH AND YMATCH        **
C     *************************************************
C
      SUBROUTINE INTEGRAL (A,B,C,D,XMATCH,YMATCH,INT)
C
      IMPLICIT NONE
C
      REAL A,B,C,D,GR,GI,RLINT,IMINT,MIDINT2D,XMATCH,YMATCH
C
      COMPLEX INT
C
      EXTERNAL GR,GI
C
      RLINT=MIDINT2D(GR,A,B,C,D,XMATCH,YMATCH)
      IMINT=MIDINT2D(GI,A,B,C,D,XMATCH,YMATCH)
C
      INT=CMPLX(RLINT,IMINT)
C
      RETURN
      END
CC
C     ***********************************
C     ** REAL PART OF GREEN'S FUNCTION **
C     ***********************************
C
      REAL FUNCTION GR(X,XPRM,Y,YPRM)
      IMPLICIT NONE
      REAL PI,K,X,XPRM,Y,YPRM,R
C
      PI=ACOS(-1.0)
      K=2.0*PI
C
      R=SQRT((X-XPRM)*(X-XPRM)+(Y-YPRM)*(Y-YPRM))
      GR=COS(K*R)/R
C
      RETURN
      END
CC
C     ****************************************
C     ** IMAGINARY PART OF GREEN'S FUNCTION **
C     ****************************************
C
      REAL FUNCTION GI(X,XPRM,Y,YPRM)
      IMPLICIT NONE
      REAL PI,K,X,XPRM,Y,YPRM,R
C
      PI=ACOS(-1.0)
      K=2.0*PI
C
      R=SQRT((X-XPRM)*(X-XPRM)+(Y-YPRM)*(Y-YPRM))
      GI=-SIN(K*R)/R
C
      RETURN
```

```
        END
CC
C
C       ******************************************************************
C       ** THIS SUBROUTINE APPROXIMATES THE INTEGRAL OF A FUNCTION  **
C       ** USING MIDPOINT INTEGRATION WITH RICHARDSON EXTRAPOLATION **
C       ******************************************************************
C
        REAL FUNCTION MIDINT2D(F,A,B,C,D,XMATCH,YMATCH)
C
        IMPLICIT NONE
        REAL CP(16,16),C1,SUMX,F,MIDINT,XMATCH,YMATCH
        REAL NSX,DX,A,B,C,D,MIDPNT,ERROR,OLD,NEW,TOL
        INTEGER COUNTX,I,Q,NMAX
        LOGICAL FINISHED
        EXTERNAL F
C
        FINISHED=.FALSE.
        COUNTX=1
        TOL=5.0E-4
        NMAX=16
C
        DO WHILE(.NOT.FINISHED)
C
C         ************************************
C         ** CALCULATE NUMBER OF SUBSECTIONS **
C         ************************************
          NSX=2.0**COUNTX
C
C         **************************
C         ** CALCULATE STEP-SIZE **
C         **************************
          DX=(B-A)/NSX
C
C         ******************************************************
C         ** EVALUATE AND SUM FUNCTION VALUE AT MIDPOINTS **
C         ******************************************************
C
          SUMX=0.0
C
          DO I=1,NSX
            MIDPNT=A+FLOAT(I-1)*DX+DX/2.0
            SUMX=SUMX+MIDINT(F,C,D,MIDPNT,XMATCH,YMATCH)
          ENDDO
C
C
C         ****************************************************
C         ** ASSIGN SUMATION TO ARRAY FOR EXTRAPOLATION **
C         ****************************************************
          CP(1,COUNTX)=SUMX*DX
C
C         ******************************************
C         ** RICHARDSON EXTRAPOLATION CALCULATION **
C         ******************************************
C
          C1=1.0
```

```
C
         DO Q=2,COUNTX
           C1=2*C1
           CP(Q,COUNTX)=(C1*CP(Q-1,COUNTX)-CP(Q-1,COUNTX-1))/(C1-1.0)
         ENDDO
C
         MIDINT2D=CP(COUNTX,COUNTX)
C
         NEW=CP(COUNTX,COUNTX)
C
C        ***************************************
C        ** CALCULATE ERROR AFTER FIRST PASS **
C        ***************************************
C
         IF((COUNTX.GT.1).AND.(NEW.NE.0.0)) ERROR=ABS(OLD-NEW)/ABS(NEW)
C
C        **************************************************************
C        ** CHECK ERROR AFTER FIRST PASS TO SEE IF INTEGRATION    **
C        ** IS WITHIN TOLERANCE END CALCULATION IF IT IS          **
C        **************************************************************
C
         IF((COUNTX.GT.1).AND.(ERRCR.LT.TOL)) FINISHED=.TRUE.
C
C        **************************************************************
C        ** CHECK TO SEE IF WE HAVE REACHED THE MAXIMUM NUMBER **
C        ** OF SUBSECTIONS ALLOWED BY USER                     **
C        **************************************************************
C
         IF(COUNTX.EQ.NMAX) FINISHED=.TRUE.
C
         COUNTX=COUNTX+1
C
         OLD=NEW
C
C
         END DO
C
         RETURN
         END
CC
C
C
C        **************************************************************
C        ** THIS SUBROUTINE APPROXIMATES THE INTEGRAL OF A FUNCTION  **
C        ** USING MIDPOINT INTEGRATION WITH RICHARDSON EXTRAPOLATION **
C        **************************************************************
C
         REAL FUNCTION MIDINT(F,A,B,X,XMATCH,YMATCH)
C
         IMPLICIT NONE
         REAL CP(16,16),C1,SUMX,F,XMATCH,YMATCH,X
         REAL NSX,DX,A,B,MIDPNT,ERROR,OLD,NEW,TOL
         INTEGER COUNTX,I,Q,NMAX
         LOGICAL FINISHED
         EXTERNAL F
```

```
C
      FINISHED=.FALSE.
      COUNTX=1
      TOL=5.0E-4
      NMAX=16
C
      DO WHILE(.NOT.FINISHED)
C
C        ************************************
C        ** CALCULATE NUMBER OF SUBSECTIONS **
C        ************************************
         NSX=2.0**COUNTX
C
C        ************************
C        ** CALCULATE STEP-SIZE **
C        ************************
         DX=(B-A)/NSX
C
C        ****************************************************
C        ** EVALUATE AND SUM FUNCTION VALUE AT MIDPOINTS **
C        ****************************************************
C
         SUMX=0.0
C
         DO I=1,NSX
           MIDPNT=A+FLOAT(I-1)*DX+DX/2.0
           SUMX=SUMX+F(X,XMATCH,MIDPNT,YMATCH)
         ENDDO
C
C        ***********************************************
C        ** ASSIGN SUMATION TO ARRAY FOR EXTRAPOLATION **
C        ***********************************************
         CP(1,COUNTX)=SUMX*DX
C
C        *********************************************
C        ** RICHARDSON EXTRAPOLATION CALCULATION **
C        *********************************************
C
         C1=1.0
C
         DO Q=2,COUNTX
           C1=2*C1
           CP(Q,COUNTX)=(C1*CP(Q-1,COUNTX)-CP(Q-1,COUNTX-1))/(C1-1.0)
         ENDDO
C
         MIDINT=CP(COUNTX,COUNTX)
C
         NEW=CP(COUNTX,COUNTX)
C
C        **************************************
C        ** CALCULATE ERROR AFTER FIRST PASS **
C        **************************************
C
         IF((COUNTX.GT.1).AND.(NEW.NE.0.0)) ERROR=ABS(OLD-NEW)/ABS(NEW)
C
```

```
C     **********************************************************
C     ** CHECK ERROR AFTER FIRST PASS TO SEE IF INTEGRATION  **
C     ** IS WITHIN TOLERANCE END CALCULATION IF IT IS        **
C     **********************************************************
C
C     IF((COUNTX.GT.1).AND.(ERROR.LT.TOL)) FINISHED=.TRUE.
C
C     **********************************************************
C     ** CHECK TO SEE IF WE HAVE REACHED THE MAXIMUM NUMBER  **
C     ** OF SUBSECTIONS ALLOWED BY USER                      **
C     **********************************************************
C
      IF(COUNTX.EQ.NMAX) FINISHED=.TRUE.
C
      COUNTX=COUNTX+1
C
      OLD=NEW
C
C
      END DO
C
      RETURN
      END
```

# Index

conductive plate
  capacitance, 24–30
  charge distribution, 30
conductive strip
  radar cross–section
    TM polarization, 61–64, 71, 73
    TE polarization, 64–69, 71, 73
  surface charge distribution
    calculated using pulses and point matching, 16–19
    calculated using Galerkin's method, 20–22
    calculation using symmetry, 22–24
    comparison of moment method solutions, 25
expansion function
  pulse, 17, 25
  rooftop, 113
  triangle, 46
FORTRAN programs
  index, 125–126
  listings, 127–252
Green's function
  two–dimensional free space, 61
  Thin wire, 36–37
Hallen's equation, 35–38
  exact kernel, 36
  moment method
  solution, 38–52
    (pulse/delta), 38–45
    (triangle/delta), 46–52
  reduced kernel
    symmetric, 37
integration
  singular integrands, 1–4

midpoint integration, 2, 7–10
  FORTRAN code, 8–9
moment method, 13–32
  use of symmetry, 22–24, 38–41
  Galerkin's method, 20–22
pulse function
  one–dimensional, 17
  two–dimensional, 25
radar cross–section (RCS)
  bistatic and monostatic, 94–97
  calculation in two–dimensions, 70–71
  conductive contour
    circular and square, 84–94
    TM case, 84–86
    TE case, 86–94
  conducting strip
    TM case, 61–64
    TE case, 64–69
  decibels
    with respect to a linear meter, $\sigma_{dBlm}$, 70
    with respect to a square meter, $\sigma_{dBsm}$, 42
  definition
    in three dimensions, 41
    in two dimensions, 70
  tapered resistive strip, 75–79
  thin plate calculation, 111–113, 117–118, 122–123
  thin wire calculation
    broadside, 41–42
    arbitrary angle, 56
resistive taper
  Taylor taper, 77–79
  quadratic taper, 75–76

Richardson's extrapolation, 4–6
  used to estimate
  derivative, 5–6
  used with midpoint
  integration, 7–10
square plate
  currents, 112, 119–120
  moment method solution (RCS)
    (pulse/pulse), 99–113
    (rooftop/pulse), 113–117
surface (sheet) resistance, 73
symmetric kernel, 23, 28, 37
symmetry
  used in moment method solution to
  find strip charge/unit length, 22–24
  used in moment method solution to
  find capaticance of plate, 28–30
  used in moment method solution to
  find RCS of thin wire, 38–41
weighting (enforcement)
function, 20–21

# The Artech House Antenna Library

Helmut E. Schrank, *Series Editor*

*Advanced Technology in Satellite Communication Antennas: Electrical and Mechanical Design*, Takashi Kitsuregawa

*Analysis Methods for Electromagnetic Wave Problems, Volume 2*, Eikichi Yamashita, ed.

*Analysis of Wire Antennas and Scatterers: Software and User's Manual*, A. R. Djordjević, M. B. Bazdar, G. M. Bazdar, G. M. Vitosevic, T. K. Sarkar, and R. F. Harrington

*Analysis Methods for Electromagnetic Wave Problems*, E. Yamashita, ed.

*Antenna-Based Signal Processing Techniques for Radar Systems*, Alfonso Farina

*Broadband Patch Antennas*, Jean-François Zürcher and Fred E. Gardiol

*CAD for Linear and Planar Antenna Arrays of Various Radiating Elements: Software and User's Manual*, Miodrag Mikavica and Aleksandar Nešić

*The CG-FFT Method: Application of Signal Processing Techniques to Electromagnetics*, Manuel F. Cátedra, Rafael P. Torres, José Basterrechea, Emilio Gago

*Electromagnetic Waves in Chiral and Bi-Isotropic Media*, I.V. Lindell, S.A. Tretyakov, A.H. Sihvola, A. J. Viitanen

*Fixed and Mobile Terminal Antennas*, A. Kumar

*Generalized Multipole Technique for Computational Electromagnetics*, Cristian Hafner

*Handbook of Antennas for EMC*, Thereza Macnamara

*Integral Equation Methods for Electromagnetics*, N. Morita, N. Kumagai, and J. Mautz

*IONOPROP: Ionospheric Propagation Assessment Program, Version 1.1: Software and User's Manual,* by Hernert V. Hitney

*Four-Armed Spiral Antennas,* Robert G. Corzine and Joseph A. Mosko

*Introduction to Electromagnetic Wave Propagation,* Paul Rohan

*Introduction to the Uniform Geometrical Theory of Diffraction,* D. A. McNamara

*Microwave Cavity Antennas,* A. Kumar and H. D. Hristov

*Millimeter-Wave Microstrip and Printed Circuit Antennas,* Prakash Bhartia

*Mobile Antenna Systems,* K. Fujimoto and J. R. James

*Modern Methods of Reflector Antenna Analysis and Design,* Craig Scott

*Moment Methods in Antennas and Scattering,* Robert C. Hansen, editor

*Monopole Elements on Circular Ground Planes,* M. M. Weiner *et al.*

*Near-Field Antenna Measurements,* D. Slater

*Passive Optical Components for Optical Fiber Transmission,* Norio Kashima

*Phased Array Antenna Handbook,* Robert J. Mailloux

*Polariztion in Electromagnetic Systems,* Warren Stutzman

*Practical Phased-Array Antenna Systems,* Eli Brookner *et al.*

*Practical Simulation of Radar Antennas and Radomes,* Herbert L. Hirsch and Douglas C. Grove

*Radiowave Propagation and Antennas for Personal Communications,* Kazimierz Siwiak

*Reflector and Lens Antennas: Software User's Manual and Example Book,* Carlyle J. Sletten, editor

*Reflector and Lens Antenna Analysis and Design Using Personal Computers,* Carlyle J. Sletten, editor

*Shipboard Antennas, Second Edition,* Preston Law

*Small-Aperture Radio Direction-Finding,* Herndon Jenkins

*Spectral Domain Method in Electromagnetics,* Craig Scott

*Understanding Electromagnetic Scattering Using the Moment Method: A Practical Approach,* Randy Bancroft

*Waveguide Components for Antenna Feed Systems: Theory and CAD,* J. Uher, J. Bornemann, and Uwe Rosenberg

*For further information on these and other Artech House titles, contact:*

Artech House
685 Canton Street
Norwood, MA 02062
617-769-9750
Fax: 617-769-6334
Telex: 951-659
email: artech@world.std.com

Artech House
Portland House - Stag Place
London SW1E 5XA England
+44 (0) 171-973-8077
Fax: +44 (0) 171-630-0166
Telex: 951-659
email: bookco@artech.demon.co.uk

www.ingramcontent.com/pod-product-compliance
Lightning Source LLC
Chambersburg PA
CBHW050128240326
41458CB00124B/1659